ILEARN

Grade 3 Math Practice

GET DIGITAL ACCESS TO

 2 ILEARN Practice Tests

 Personalized Study Plans

REGISTER NOW

Link — **QR Code**

Visit the link below for online registration

lumoslearning.com/a/tedbooks

Access Code: G3MILEARN-59280-P

ILEARN Test Prep: 3rd Grade Math Practice Workbook and Full-length Online Assessments: Indiana Learning Evaluation Assessment Readiness Network Study Guide

Contributing Editor - Keyana M. Martinez
Contributing Editor - LaSina McLain-Jackson
Contributing Editor - Greg Applegate
Executive Producer - Mukunda Krishnaswamy
Program Director - Anirudh Agarwal
Designer and Illustrator - Nagendra K V

COPYRIGHT ©2018 by Lumos Information Services, LLC. ALL RIGHTS RESERVED. No portion of this book may be reproduced mechanically, electronically or by any other means, including photocopying, recording, taping, Web Distribution or Information Storage and Retrieval systems, without prior written permission of the Publisher, Lumos Information Services, LLC.

Indiana Department of Education is not affiliated to Lumos Learning. Indiana Department of Education, was not involved in the production of, and does not endorse these products or this site.

ISBN 10: 1946795704

ISBN 13: 978-1946795700

Printed in the United States of America

FOR SCHOOL EDITION AND PERMISSIONS, CONTACT US

LUMOS INFORMATION SERVICES, LLC

PO Box 1575, Piscataway, NJ 08855-1575
www.LumosLearning.com

Email: support@lumoslearning.com
Tel: (732) 384-0146
Fax: (866) 283-6471

INTRODUCTION

This book is specifically designed to improve student achievement on the Indiana's Learning Evaluation and Assessment Readiness Network (ILEARN). Students perform at their best on standardized tests when they feel comfortable with the test content as well as the test format. Lumos online practice tests are meticulously designed to mirror the state assessment. They adhere to the guidelines provided by the state for the number of sessions and questions, standards, difficulty level, question types, test duration and more.

Based on our decade of experience developing practice resources for standardized tests, we've created a dynamic system, the Lumos Smart Test Prep Methodology. It provides students with realistic assessment rehearsal and an efficient pathway to overcoming each proficiency gap.

Use the Lumos Smart Test Prep Methodology to achieve a high score on the ILEARN.

Lumos Smart Test Prep Methodology

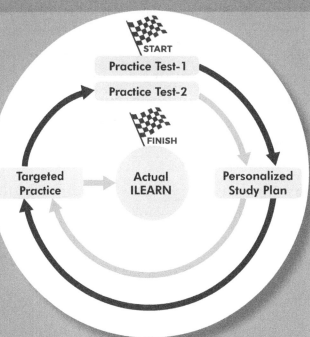

1 The student takes the first online diagnostic test, which assesses proficiency levels in various standards.

2 StepUp generates a personalized online study plan based on the student's performance.

3 The student completes targeted practice in the printed workbook and marks it as complete in the online study plan.

4 The student then attempts the second online practice test.

5 StepUp generates a second individualized online study plan.

6 The student completes the targeted practice and is ready for the actual ILEARN.

Table of Contents

Introduction			1
Chapter 1	**Number Sense**		**4**
Lesson 1	3.NS.1	Read and write numbers to 1000 using base-ten numerals	4
Lesson 2	3.NS.2	Fractions of a Whole	6
Lesson 3	3.NS.3	Fractions on the Number Line	13
Lesson 4	3.NS.4 & 5	Comparing Fractions	20
Lesson 5	3.NS.6	Rounding Numbers	26
		Answer Key & Detailed Explanations	30
Chapter 2	**Computation and Algebraic Thinking**		**43**
Lesson 1	3.CA.1	Addition & Subtraction	43
Lesson 2	3.CA.2	Two-Step Problems	47
Lesson 3	3.CA.3	Understanding Multiplication	52
Lesson 4	3.CA.4	Understanding Division	60
Lesson 5	3.CA.5	Multiplication & Division Facts	65
Lesson 6	3.CA.6	Multiplication & Division Properties	72
Lesson 7	3.CA.7	Applying Multiplication & Division	77
Lesson 8	3.CA.8	Number Patterns	82
		Answer Key & Detailed Explanations	87
Chapter 3	**Geometry**		**105**
Lesson 1	3.G.1	2-Dimensional Shapes	105
Lesson 2	3.G.3	Shape Partitions	110
		Answer Key & Detailed Explanations	118

Chapter 4	**Measurement**		....	**123**
Lesson 1	3.M.1	Liquid Volume & Mass	....	123
Lesson 2	3.M.2	Measuring Length	....	127
Lesson 3	3.M.3	Telling Time	....	131
Lesson 4	3.M.3	Elapsed Time	....	137
Lesson 5	3.M.4	Solve word problems involving money	....	142
Lesson 6	3.M.5	Area	....	144
Lesson 7	3.M.5	Relating Area to Addition & Multiplication	....	149
Lesson 8	3.M.6	Perimeter	....	153
		Answer Key & Detailed Explanations	....	**162**

Chapter 5	**Data Analysis**		....	**180**
Lesson 1	3.DA.1	2-Dimensional Shapes	....	180
		Answer Key & Detailed Explanations	....	**191**

Additional Information	....	**195**
Test Taking Tips and FAQs	....	195
What if I buy more than one Lumos tedBook?	....	196
Progress Chart	....	197

Date of Completion:_____   Score:_____

Chapter 1: Number Sense

Lesson 1: Read and write numbers to 1000 using base-ten numerals

1. How is 458 written in expanded form?

   Ⓐ 400 + 500 + 800
   Ⓑ 45 + 8
   Ⓒ 400 + 50 + 8
   Ⓓ 400 + 58

2. Which number is five hundred thirty?

   Ⓐ 503
   Ⓑ 530
   Ⓒ 533
   Ⓓ 513

3. What is 200 + 8 written in standard form?

   Ⓐ 280
   Ⓑ 288
   Ⓒ 208
   Ⓓ 2008

4. What is 167 written in word form?

   Ⓐ One hundred sixty-seven
   Ⓑ One hundred six hundred seven
   Ⓒ One hundred six seventy
   Ⓓ One hundred sixty

5. Select the correct way to write 734 in word and expanded form.

   Ⓐ 700 + 34
   Ⓑ Seven hundred thirty-four
   Ⓒ Seventy thirty-four
   Ⓓ 700 + 30 + 4
   Ⓔ 730 + 4

4

LumosLearning.com

6. Select the correct way to write 480 in word and expanded form.

   Ⓐ Four hundred eight
   Ⓑ Four hundred eighty-eight
   Ⓒ Four hundred eighty
   Ⓓ 400+80+0
   Ⓔ 480+0

7. What number is expressed as 500 + 2 in expanded form?

   [   ]

8. Match the table with the correct number.

	437	430	473	407
Four hundred seven				
400 + 30				
Four hundred seventy-three				
400 +30 +7				

9. Complete the table by filling in the different forms under each column.

STANDARD	EXPANDED	WORD
444	400 + 40 + 4	
	300 + 8	Three hundred eight
		Nine hundred twenty five
	400 + 10	

10. Mya says 300 when written in expanded form is 300. Trevor says that 300 written in expanded form is 300 + 0 + 0. Who is correct? Explain your reasoning.

Date of Completion:_____ Score:_____

CHAPTER 1 → Lesson 2: Fractions of a Whole

1. What fraction of the letters in the word "READING" are vowels?

    Ⓐ $\frac{4}{7}$

    Ⓑ $\frac{3}{4}$

    Ⓒ $\frac{3}{7}$

    Ⓓ $\frac{1}{3}$

2. A bag contains 3 red, 2 yellow, and 5 blue tiles. What fraction of the tiles are yellow?

    Ⓐ $\frac{2}{5}$

    Ⓑ $\frac{2}{10}$

    Ⓒ $\frac{3}{7}$

    Ⓓ $\frac{1}{3}$

3. A rectangle is cut into four equal pieces. Each piece represents what fraction of the rectangle?

    Ⓐ one half
    Ⓑ one third
    Ⓒ one fourth
    Ⓓ one fifth

4. What fraction of the square is shaded?

Ⓐ $\dfrac{1}{2}$ Ⓒ $\dfrac{2}{1}$

Ⓑ $\dfrac{1}{3}$ Ⓓ $\dfrac{1}{1}$

5. What fraction of the square is shaded?

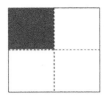

Ⓐ $\dfrac{1}{2}$

Ⓑ $\dfrac{1}{4}$

Ⓒ $\dfrac{1}{3}$

Ⓓ $\dfrac{3}{1}$

6. What fraction of the square is NOT shaded?

Ⓐ $\dfrac{1}{2}$

Ⓑ $\dfrac{1}{4}$

Ⓒ $\dfrac{3}{1}$

Ⓓ $\dfrac{3}{4}$

7. What fraction of the circle is shaded?

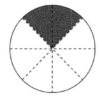

Ⓐ $\dfrac{1}{8}$

Ⓑ $\dfrac{2}{8}$

Ⓒ $\dfrac{2}{6}$

Ⓓ $\dfrac{6}{2}$

8. What fraction of the circle is not shaded?

Ⓐ $\dfrac{6}{8}$

Ⓑ $\dfrac{7}{8}$

Ⓒ $\dfrac{2}{6}$

Ⓓ $\dfrac{6}{2}$

9. What fraction of the circle is shaded?

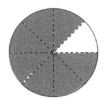

Ⓐ $\dfrac{1}{8}$

Ⓑ $\dfrac{1}{7}$

Ⓒ $\dfrac{7}{1}$

Ⓓ $\dfrac{7}{8}$

10. What fraction of the circle is not shaded?

Ⓐ $\frac{1}{8}$

Ⓑ $\frac{1}{7}$

Ⓒ $\frac{7}{1}$

Ⓓ $\frac{8}{1}$

11. What fraction of the rectangle is shaded?

Ⓐ $\frac{1}{2}$

Ⓑ $\frac{1}{3}$

Ⓒ $\frac{2}{3}$

Ⓓ $\frac{2}{1}$

12. What fraction of the rectangle is not shaded?

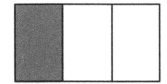

Ⓐ $\frac{1}{3}$

Ⓑ $\frac{1}{2}$

Ⓒ $\frac{2}{3}$

Ⓓ $\frac{2}{1}$

13. A pizza is cut into 12 equal slices. Eight slices are eaten. What fraction of the pizza is left?

Ⓐ $\frac{8}{12}$

Ⓑ $\frac{4}{8}$

Ⓒ $\frac{4}{12}$

Ⓓ $\frac{8}{4}$

14. The class has 20 children. Only half of the students brought their homework. How many students have their homework?

Ⓐ 20 students
Ⓑ 15 students
Ⓒ 10 students
Ⓓ 12 students

15. Meagan has 24 cupcakes. She gives a third of them to Micah. How many cupcakes does Micah have?

Ⓐ 8 cupcakes
Ⓑ 12 cupcakes
Ⓒ 3 cupcakes
Ⓓ 4 cupcakes

16. Which of the following fractions could apply to this figure? Complete the table by selecting yes or no.

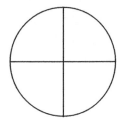

	Yes	No
1/8	○	○
1/4	○	○
1/3	○	○

17. What fraction does each figure show? Write your answers in the blank boxes.

Figure	Fraction

18. Which of the following fractions could apply to this figure? Select all correct answers.

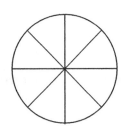

Ⓐ $\dfrac{1}{3}$

Ⓑ $\dfrac{1}{8}$

Ⓒ $\dfrac{1}{5}$

Ⓓ $\dfrac{8}{8}$

CHAPTER 1 → Lesson 3: Fractions on the Number Line

1. What fraction does the number line show?

   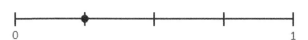

   Ⓐ $\dfrac{1}{4}$

   Ⓑ $\dfrac{1}{3}$

   Ⓒ $\dfrac{3}{4}$

   Ⓓ $\dfrac{4}{4}$

2. What fraction does the number line show?

   Ⓐ $\dfrac{1}{2}$

   Ⓑ $\dfrac{2}{2}$

   Ⓒ $\dfrac{1}{3}$

   Ⓓ $\dfrac{2}{3}$

3. What fraction does the number line show?

   Ⓐ $\dfrac{2}{8}$

   Ⓑ $\dfrac{3}{5}$

   Ⓒ $\dfrac{3}{8}$

   Ⓓ $\dfrac{4}{8}$

4. What fraction does the number line show?

   Ⓐ $\dfrac{3}{8}$

   Ⓑ $\dfrac{6}{8}$

   Ⓒ $\dfrac{5}{8}$

   Ⓓ $\dfrac{4}{8}$

5. What fraction does the number line show?

   Ⓐ $\dfrac{1}{6}$

   Ⓑ $\dfrac{4}{6}$

   Ⓒ $\dfrac{3}{6}$

   Ⓓ $\dfrac{1}{5}$

6. What fraction does the number line show?

Ⓐ $\frac{2}{4}$

Ⓑ $\frac{2}{3}$

Ⓒ $\frac{1}{3}$

Ⓓ $\frac{1}{4}$

7. What fraction does the number line show?

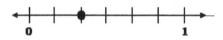

Ⓐ $\frac{1}{6}$

Ⓑ $\frac{3}{6}$

Ⓒ $\frac{2}{4}$

Ⓓ $\frac{2}{6}$

8. What fraction does the number line show?

Ⓐ $\frac{4}{8}$

Ⓑ $\frac{5}{8}$

Ⓒ $\frac{4}{4}$

Ⓓ $\frac{2}{8}$

9. What fraction does the number line show?

Ⓐ $\frac{2}{8}$

Ⓑ $\frac{2}{9}$

Ⓒ $\frac{1}{9}$

Ⓓ $\frac{1}{8}$

10. What fraction does the number line show

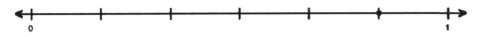

Ⓐ $\frac{4}{6}$

Ⓑ $\frac{5}{6}$

Ⓒ $\frac{3}{6}$

Ⓓ $\frac{1}{6}$

11. What fraction does the number line show?

Ⓐ $\frac{2}{3}$

Ⓑ $\frac{1}{3}$

Ⓒ $\frac{3}{3}$

Ⓓ $\frac{4}{3}$

12. What fraction does the number line show?

- Ⓐ $\frac{1}{4}$
- Ⓑ $\frac{2}{4}$
- Ⓒ $\frac{3}{4}$
- Ⓓ $\frac{4}{4}$

13. What fraction does the number line show?

- Ⓐ $\frac{8}{9}$
- Ⓑ $\frac{7}{8}$
- Ⓒ $\frac{2}{8}$
- Ⓓ $\frac{5}{9}$

14. What fraction does the number line show?

- Ⓐ $\frac{1}{4}$
- Ⓑ $\frac{3}{3}$
- Ⓒ $\frac{2}{4}$
- Ⓓ $\frac{3}{4}$

15. What fraction does the number line show?

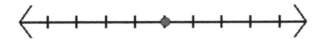

Ⓐ $\frac{7}{8}$

Ⓑ $\frac{6}{8}$

Ⓒ $\frac{5}{8}$

Ⓓ $\frac{5}{3}$

16. Which fractions does the number line show? Select all correct answers.

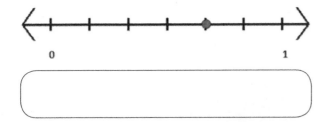

Ⓐ $\frac{3}{6}$

Ⓑ $\frac{1}{4}$

Ⓒ $\frac{4}{8}$

Ⓓ $\frac{1}{2}$

17. What fraction does the number line show? Write your answer in the box below.

18. Draw a number line and locate the fraction $\frac{5}{7}$ on it.

19. There are 3 number lines in the first column. Which fractions are represented by the dots on the number lines? For each number line, select the correct answer.

Date of Completion:_____   Score:_____

CHAPTER 1 → Lesson 4: Comparing Fractions

1. Which of these sets has the fractions listed from least to greatest?

   Ⓐ $\frac{1}{6}, \frac{1}{4}, \frac{1}{3}, \frac{1}{2}$

   Ⓑ $\frac{1}{2}, \frac{1}{3}, \frac{1}{6}, \frac{1}{4}$

   Ⓒ $\frac{1}{3}, \frac{1}{4}, \frac{1}{2}, \frac{1}{6}$

   Ⓓ $\frac{1}{2}, \frac{1}{3}, \frac{1}{4}, \frac{1}{6}$

2. Which of these fractions would be found between $\frac{1}{2}$ and 1 on a number line?

   Ⓐ $\frac{1}{4}$

   Ⓑ $\frac{1}{3}$

   Ⓒ $\frac{5}{8}$

   Ⓓ $\frac{3}{1}$

3. Which of these fractions would be found between 0 an $\frac{1}{2}$ on a number line?

   Ⓐ $\frac{7}{8}$

   Ⓑ $\frac{3}{4}$

   Ⓒ $\frac{1}{4}$

   Ⓓ $\frac{5}{8}$

4. Which of these fractions would be found between 0 and $\frac{3}{4}$ on a number line?

    Ⓐ $\frac{7}{8}$

    Ⓑ $\frac{4}{8}$

    Ⓒ $\frac{5}{6}$

    Ⓓ $\frac{4}{4}$

5. Which of these fractions is less than $\frac{6}{8}$?

    Ⓐ $\frac{1}{8}$

    Ⓑ $\frac{7}{8}$

    Ⓒ $\frac{9}{8}$

    Ⓓ $\frac{8}{8}$

6. Answer the following: $\frac{1}{2}$ > _____?

    Ⓐ $\frac{1}{4}$

    Ⓑ $\frac{2}{3}$

    Ⓒ $\frac{4}{8}$

    Ⓓ $\frac{2}{2}$

7. Which is greater: $\frac{4}{8}$ or $\frac{1}{2}$?

    Ⓐ $\frac{1}{2}$

    Ⓑ $\frac{4}{8}$

    Ⓒ They are equal.

    Ⓓ There is not enough information given.

8. Which fraction is less: $\frac{1}{4}$ or $\frac{1}{8}$?

   Ⓐ $\frac{1}{4}$

   Ⓑ $\frac{1}{8}$

   Ⓒ They are equal.

   Ⓓ There is not enough information given.

9. Which fraction is less $\frac{4}{6}$ or $\frac{1}{6}$?

   Ⓐ $\frac{4}{6}$

   Ⓑ $\frac{1}{6}$

   Ⓒ They are equal.

   Ⓓ There is not enough information given.

10. If two fractions have the same denominator, the one with a(n) _____ numerator is the greater fraction.

    Ⓐ smaller
    Ⓑ greater
    Ⓒ even
    Ⓓ zero

11. Complete this number sentence: $\frac{4}{6}$ _____ $\frac{5}{6}$

    Ⓐ >
    Ⓑ <
    Ⓒ =
    Ⓓ There is not enough information given.

12. Complete this number sentence: $\frac{3}{8}$ _____ $\frac{5}{8}$

    Ⓐ >
    Ⓑ <
    Ⓒ =
    Ⓓ There is not enough information given.

13. Complete this number sentence: $\frac{2}{4}$ _____ $\frac{1}{2}$

    Ⓐ >
    Ⓑ <
    Ⓒ =
    Ⓓ There is not enough information given.

14. Complete this number sentence: $\frac{1}{2}$ _____ $\frac{1}{3}$

    Ⓐ >
    Ⓑ <
    Ⓒ =
    Ⓓ There is not enough information given.

15. To compare fractions with the same numerator, you need to look at the _____.

    Ⓐ numerators
    Ⓑ denominators
    Ⓒ factors of the numerator
    Ⓓ multiples of the denominator

16. Are the following fractions less than $\frac{3}{4}$? Select yes or no.

	Yes	No
$\frac{1}{4}$	○	○
$\frac{4}{4}$	○	○
$\frac{2}{4}$	○	○

17. Is Fraction #1 less than, greater than, or equal to Fraction #2? Write the correct symbol in the empty boxes.

Fraction #1	< = or >	Fraction #2
$\frac{2}{3}$		$\frac{1}{3}$
$\frac{4}{6}$		$\frac{5}{6}$
$\frac{8}{8}$		$\frac{5}{5}$
$\frac{3}{4}$		$\frac{1}{4}$

18. Which of the following fractions are greater than $\frac{2}{5}$? Select all correct answers.

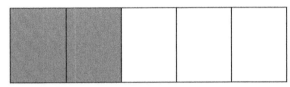

Ⓐ $\frac{1}{5}$

Ⓑ $\frac{3}{5}$

Ⓒ $\frac{4}{5}$

Ⓓ $\frac{2}{5}$

19. Which of these fractions is greater than $\frac{5}{7}$? Circle the correct answer.

Ⓐ $\frac{1}{7}$

Ⓑ $\frac{2}{7}$

Ⓒ $\frac{4}{7}$

Ⓓ $\frac{6}{7}$

20. Which of the following fractions is the least?

$\frac{6}{5}, \frac{6}{4}, \frac{6}{10}, \frac{6}{9}.$

Write your answer in the box below.

Date of Completion:_____ Score:_____

CHAPTER 1 → Lesson 5: Rounding Numbers

1. What is the value of the 9 in 11,291?

   Ⓐ 9 ones
   Ⓑ 9 hundreds
   Ⓒ 9 thousands
   Ⓓ 9 tens

2. What is the value of the digit 6 in 36,801?

   Ⓐ Six thousand
   Ⓑ Sixty
   Ⓒ Sixty thousand
   Ⓓ Six hundred

3. Which of these numbers has a 9 in the thousands place?

   Ⓐ 690,099
   Ⓑ 900
   Ⓒ 209,866
   Ⓓ 90,786

4. Round 2,564 to the nearest hundred.

   Ⓐ 2,000
   Ⓑ 2,500
   Ⓒ 2,600
   Ⓓ 2,700

5. Round 1,043 to the nearest hundred.

   Ⓐ 1,000
   Ⓑ 1,100
   Ⓒ 1,040
   Ⓓ 1,200

6. Round 537 to the nearest ten.

    Ⓐ 500
    Ⓑ 540
    Ⓒ 550
    Ⓓ 530

7. Round 957 to the nearest ten.

    Ⓐ 960
    Ⓑ 950
    Ⓒ 900
    Ⓓ 1,000

8. Maya is buying pencils for the school. Maya needs to buy enough pencils for 388 students. What is this number rounded to the nearest hundred?

    Ⓐ 390
    Ⓑ 380
    Ⓒ 400
    Ⓓ 500

9. Ninety-seven chairs are needed for an audience. What is this number rounded to the nearest ten?

    Ⓐ 90
    Ⓑ 100
    Ⓒ 80
    Ⓓ 110

10. Which of the following numbers does not round to 1,000 when rounding to the nearest hundred?

    Ⓐ 955
    Ⓑ 1,005
    Ⓒ 1,051
    Ⓓ 951

11. How many whole numbers, when rounded to the nearest ten give 100 as the result?

    Ⓐ 8
    Ⓑ 9
    Ⓒ 10
    Ⓓ 11

12. Fill in the blank.
    795 rounds to 800 when rounded to the nearest _____.

    Ⓐ ten
    Ⓑ hundred
    Ⓒ ten or hundred
    Ⓓ thousand

13. Fill in the blank.
    1,090 rounds to 1,100 when rounded to the nearest _____.

    Ⓐ ten
    Ⓑ hundred
    Ⓒ ten or hundred
    Ⓓ thousand

14. The attendance at a local baseball game is announced to be 4,328. What is this number rounded to the nearest ten?

    Ⓐ 4,300
    Ⓑ 4,330
    Ⓒ 4,320
    Ⓓ 4,400

15. The number of plants in a garden, when rounded to the nearest hundred, rounds to 800. Which of the following could not be the number of plants in the garden?

    Ⓐ 850
    Ⓑ 800
    Ⓒ 750
    Ⓓ 849

16. Which numbers represent the number 617 when rounded to the nearest ten or hundred? Circle all correct answers.

    Ⓐ 620
    Ⓑ 600
    Ⓒ 700
    Ⓓ 630

17. Round 489 to the nearest hundred. Write the correct answer into the box.

18. Complete the table in the format given in the example.

Number	Number when rounded to the nearest ten	Number when rounded to the nearest hundred
2,349	2,350	2,300
4,092		
8,396		

End of Number Sense

ANSWER KEY AND DETAILED EXPLANATION

Chapter 1: Number Sense

Lesson 1: Read And Write Numbers To 1000 Using Base-ten Numerals

Question No.	Answer	Detailed Explanations
1	C	The answer is C. When writing a number in expanded form you add the value of each number. A 4 in the hundreds place has a value of 400. A 5 in the tens place has a value of 50. A 8 in the ones place has a value of 8.
2	B	The answer is B. 530.
3	C	The answer is C. When writing a number in expanded form you add the value of each number. A 2 in the hundreds place has a value of 200. A 8 in the ones place has a value of 8.
4	A	The answer is A. When writing a number, you write the number how you say it. It is important that students remember to not say "and" when saying whole numbers.
5	B & D	The answers are B & D. When writing a number in expanded form you add the value of each digit. When writing a number in word form, you write the number how you say it.
6	C & D	The answers are C & D. When writing a number in expanded form you add the value of each digit. When writing a number in word form, you write the number how you say it.
7	502	502
8		Four hundred seven --> 407 400 + 30 --> 430 Four hundred seventy-three --> 473 400 +30 +7 --> 437

Question No.	Answer	Detailed Explanations		
9		STANDARD	EXPANDED	WORD
		444	400 + 40 + 4	**Four hundred forty four**
		308	300 + 8	Three hundred eight
		925	**900 + 20 + 5**	Nine hundred twenty five
		410	400 + 10	**Four hundred ten**
10		Mya is correct. When writing a number in expanded form you add the values of each digit. You do not include 0 in expanded form. Even though Trevor's answer would still equal 300, that is not the correct way to write 300 in expanded form.		

Lesson 2: Fractions of a Whole

Question No.	Answer	Detailed Explanation
1	C	When forming a fraction, the numerator will be the part of the whole and the denominator will be the whole or all parts together. In this case, there are 3 vowels (the part) and there are 7 total letters (the whole). The fraction should be $\frac{3}{7}$.
2	B	When forming a fraction, the numerator will be the part of the whole and the denominator will be the whole or all parts together. In this case, there are 2 yellow tiles (the part) and there are 10 total tiles (the whole). The fraction should be $\frac{2}{10}$.
3	C	When forming a fraction, the numerator will be the part of the whole and the denominator will be the whole or all parts together. In this case, there is 1 piece (the part) and there are 4 pieces (the whole). The fraction should be $\frac{1}{4}$.
4	A	When forming a fraction, the numerator will be the part of the whole and the denominator will be the whole or all parts together. In this case, there is 1 shaded part (the part) and there are 2 total parts (the whole). The fraction should be $\frac{1}{2}$.
5	B	When forming a fraction, the numerator will be the part of the whole and the denominator will be the whole or all parts together. In this case, there is 1 shaded part (the part) and there are 4 total parts (the whole). The fraction should be $\frac{1}{4}$.
6	D	When forming a fraction, the numerator will be the part of the whole and the denominator will be the whole or all parts together. In this case, there are 3 NOT shaded parts (the part) and there are 4 total parts (the whole). The fraction should be $\frac{3}{4}$.

Question No.	Answer	Detailed Explanation
7	B	When forming a fraction, the numerator will be the part of the whole and the denominator will be the whole or all parts together. In this case, there are 2 shaded parts (the part) and there are 8 total parts (the whole). The fraction should be $\frac{2}{8}$.
8	A	When forming a fraction, the numerator will be the part of the whole and the denominator will be the whole or all parts together. In this case, there are 6 NOT shaded parts (the part) and there are 8 total parts (the whole). The fraction should be $\frac{6}{8}$.
9	D	When forming a fraction, the numerator will be the part of the whole and the denominator will be the whole or all parts together. In this case, there are 7 shaded parts (the part) and there are 8 total parts (the whole). The fraction should be $\frac{7}{8}$.
10	A	When forming a fraction, the numerator will be the part of the whole and the denominator will be the whole or all parts together. In this case, there is 1 NOT shaded part (the part) and there are 8 total parts (the whole). The fraction should be $\frac{1}{8}$.
11	B	When forming a fraction, the numerator will be the part of the whole and the denominator will be the whole or all parts together. In this case, there is 1 shaded part (the part) and there are 3 total parts (the whole). The fraction should be $\frac{1}{3}$.
12	C	When forming a fraction, the numerator will be the part of the whole and the denominator will be the whole or all parts together. In this case, there are 2 NOT shaded parts (the part) and there are 3 total parts (the whole). The fraction should be $\frac{2}{3}$.
13	C	When forming a fraction, the numerator will be the part of the whole and the denominator will be the whole or all parts together. In this case, you subtract the slices eaten from the total number of slices (12-8) to find that there are 4 slices left (the part). Since there were 12 total slices (the whole), the fraction should be $\frac{4}{12}$.
14	C	Half is equivalent to dividing a number by 2 and 20 ÷ 2 = 10.
15	A	A third is equivalent to dividing a number by 3 and 24 ÷ 3 = 8.

Question No.	Answer	Detailed Explanation
16		

	Yes	No
1/8	○	●
1/4	●	○
1/3	○	●

The circle is divided into 4 equal parts. 1 of the 4 parts represents ¼ of the whole circle.

Question No.	Answer	Detailed Explanation
17		

	Figure	Fraction
A		$\frac{1}{8}$
B		$\frac{1}{5}$
C		$\frac{1}{2}$

The correct answers are $\frac{1}{8}$, $\frac{1}{5}$, and $\frac{1}{2}$. Figure A is divided into 8 equal parts. 1 of the 8 parts represents $\frac{1}{8}$ of the whole figure. Figure B is divided into 5 equal parts. 1 of the 5 parts represents $\frac{1}{5}$ of the whole figure. Figure C is divided into 2 equal parts. 1 of the 2 parts represents $\frac{1}{2}$ of the whole figure.

| 18 | B & D | The circle is divided into 8 equal parts. 1 of the 8 parts represents $\frac{1}{8}$ of the whole circle. The fraction $\frac{8}{8}$ represents the whole circle. $\frac{8}{8}=1$. |

Lesson 3: Fractions on the Number Line

Question No.	Answer	Detailed Explanation
1	A	The number line is divided into four segments and the dot is at the first segment of the four. The fraction is $\frac{1}{4}$.
2	A	The number line is divided into two segments and the dot is at the first segment of the two. The fraction is $\frac{1}{2}$.
3	C	The number line is divided into eight segments and the dot is at the third segment of the eight. The fraction is $\frac{3}{8}$.
4	B	The number line is divided into eight segments and the dot is at the sixth segment of the eight. The fraction is $\frac{6}{8}$.
5	A	The number line is divided into six segments and the dot is at the first segment of the six. The fraction is $\frac{1}{6}$.
6	C	The number line is divided into three segments and the dot is at the first segment of the three. The fraction is $\frac{1}{3}$.
7	D	The number line is divided into six segments and the dot is at the second segment of the six. The fraction is $\frac{2}{6}$.
8	A	The number line is divided into eight segments and the dot is at the fourth segment of the eight. The fraction is $\frac{4}{8}$.
9	D	The number line is divided into eight segments and the dot is at the first segment of the eight. The fraction is $\frac{1}{8}$.
10	B	The number line is divided into six segments and the dot is at the fifth segment of the six. The fraction is $\frac{5}{6}$.
11	A	The number line is divided into three segments and the dot is at the second segment of the three. The fraction is $\frac{2}{3}$.
12	B	The number line is divided into four segments and the dot is at the second segment of the four. The fraction is $\frac{2}{4}$.
13	B	The number line is divided into eight segments and the dot is at the seventh segment of the eight. The fraction is $\frac{7}{8}$.
14	D	The number line is divided into four segments and the dot is at the third segment of the four. The fraction is $\frac{3}{4}$.
15	C	The number line is divided into eight segments and the dot is at the fifth segment of the eight. The fraction is $\frac{5}{8}$.

Question No.	Answer	Detailed Explanation
16	C & D	The number line from 0 to 1 is divided into 8 equal segments. The number line is marked at the 4th segment. This represents $\frac{4}{8}$ of the whole line. $\frac{4}{8}$ can also be seen as $\frac{1}{2}$.
17	$\frac{4}{6}$	The number line from 0 to 1 is divided into 6 equal segments. The number line is marked at the 4th segment. This represents $\frac{4}{6}$ of the whole line.
18		The interval from 0 to 1 is taken as the whole and is divided into 7 equal segments. Each segment has a size $\frac{1}{7}$. $\frac{5}{7}$ is represented by the Red dot located at the end of the 5th segment from 0 as shown in the figure below. (0/7) 0 1/7 2/7 3/7 4/7 5/7 6/7 1 (= 7/7)
19		

	4/4	8/9	5/8
(number line 1)	○	●	○
(number line 2)	○	○	●
(number line 3)	●	○	○

In the first number line, the interval from 0 to 1 is taken as the whole and is divided into 9 equal segments. Each segment has a size $\frac{1}{9}$. The dot is at the end of the 8th segment from 0. So, it represents $\frac{8}{9}$.

In the second number line, the interval from 0 to 1 is taken as the whole and is divided into 8 equal segments. Each segment has a size $\frac{1}{8}$. The dot is at the end of the 5th segment from 0. So, it represents $\frac{5}{8}$.

In the third number line, the interval from 0 to 1 is taken as the whole and is divided into 4 equal segments. Each segment has a size $\frac{1}{4}$. The dot is at the end of the 4th segment from 0. So, it represents $\frac{4}{4}$ or 1.

Lesson 4: Comparing Fractions

Question No.	Answer	Detailed Explanation
1	A	If the numerators of two or more fractions are the same, the fraction with the greatest denominator is the smallest fraction. Option A is the only choice that has the numbers lined up from the greatest to the smallest denominator.
2	C	In order for a fraction to be between $\frac{1}{2}$ and 1, it would have to be greater than $\frac{1}{2}$. Option C is the only choice that fits this criteria. Since $\frac{4}{8} = \frac{1}{2}$, $\frac{5}{8}$ would be greater than $\frac{1}{2}$. $\frac{3}{1}$ is greater than 1 whole, so that is too large. $\frac{1}{3}$ is smaller than $\frac{1}{2}$.
3	C	In order for a fraction to be between 0 and 1/2, it would have to be less than $\frac{1}{2}$. Option C is the only choice that fits this criteria.
4	B	In order for a fraction to be between 0 and $\frac{3}{4}$, it would have to be less than $\frac{3}{4}$. Option B is the only choice that fits this criteria. (Note: $\frac{4}{8}$ is equivalent to $\frac{1}{2}$.)
5	A	When comparing fractions, if the denominators are the same, compare the numerators to see which one is the smallest.
6	A	The answer has to be less than $\frac{1}{2}$. Option A is the only choice that fits this criteria. $\frac{1}{4}$ is less than $\frac{1}{2}$, since fourths are smaller parts of a whole than halves. $\frac{4}{8}$ is equal to $\frac{1}{2}$, and $\frac{3}{4}$ is greater.
7	C	The fraction $\frac{4}{8}$ is at the same point on a number line as $\frac{1}{2}$; therefore the two fractions are equal. When the numerator and denominator of $\frac{4}{8}$ are each divided by 4, the fraction can be simplified to $\frac{1}{2}$.
8	B	If the numerators of two fractions are the same, the fraction with the greater denominator is the smaller fraction. In this case, $\frac{1}{8}$ is the smaller fraction.
9	B	When comparing fractions and the denominators are the same, compare the numerators. In this case, 1 is less than 4. In this case $\frac{1}{4}$ is the smaller fraction.
10	B	When comparing fractions, if the denominators are the same, compare the numerators to see which one is the largest fractions. When comparing $\frac{2}{9}$ and $\frac{4}{9}$, $\frac{4}{9}$ would be greater because it has a greater numerator.

Question No.	Answer	Detailed Explanation
11	B	When comparing fractions where the denominators are the same, compare the numerators. In this case, 4 is less than 5. So $\frac{4}{6} < \frac{5}{6}$ or $\frac{4}{6}$ is the smaller fraction.
12	B	When comparing fractions where the denominators are the same, compare the numerators. In this case, 3 is less than 5. So $\frac{3}{8}$ is the smaller fraction. So $\frac{3}{8} < \frac{5}{8}$.
13	C	The fraction $\frac{2}{4}$ is at the same point on a number line as $\frac{1}{2}$; therefore the two fractions are equal.
14	A	If the numerators of two fractions are the same, the fraction with the lesser denominator is the greater fraction. In this case, $\frac{1}{2}$ is greater than $\frac{1}{2}$.
15	B.	If the numerators of two fractions are the same, compare the denominators. The fraction with the greater denominator is the smaller fraction.
16		

	Yes	No
$\frac{1}{4}$	●	
$\frac{4}{4}$		●
$\frac{2}{4}$	●	

Fractions which refer to the same denominator can be compared in size. The fraction $\frac{1}{4}$ represents a smaller portion of the whole than $\frac{3}{4}$. Therefore, $\frac{1}{4} < \frac{3}{4}$. The fraction $\frac{4}{4}$ represents the entire whole or 1. Therefore, $\frac{4}{4} > \frac{3}{4}$. The fraction $\frac{2}{4}$ or $\frac{1}{2}$ represents a smaller portion of the whole than $\frac{3}{4}$. Therefore, $\frac{2}{4} < \frac{3}{4}$.

Lesson 5: Rounding Numbers

Question No.	Answer	Detailed Explanation
1	D	Moving from right to left, the positions are as follows: ones, tens, hundreds, thousands, ten thousands. 9 - 10's is the same as 9 x 10 = 90.
2	A	Moving from right to left, the positions are as follows: ones, tens, hundreds, thousands, ten thousands.
3	C	Moving from right to left, the positions are as follows: ones, tens, hundreds, thousands, ten thousands.
4	C	Moving from right to left, the positions are as follows: ones, tens, hundreds, thousands. In order to round to the nearest hundred, you must look at the number in the tens place. If this number is less than 5, you must round the hundreds number down. If this number is 5 or more, you must round the hundreds number up.
5	A	Moving from right to left, the positions are as follows: ones, tens, hundreds, thousands. In order to round to the nearest hundred, you must look at the number in the tens place. If this number is less than 5, you must round the hundreds number down. If this number is 5 or more, you must round the hundreds number up.
6	B	Moving from right to left, the positions are as follows: ones, tens, hundreds. In order to round to the nearest ten, you must look at the number in the ones place. If this number is less than 5, you must round the tens number down. If this number is 5 or more, you must round the tens number up.
7	A	Moving from right to left, the positions are as follows: ones, tens, hundreds. In order to round to the nearest ten, you must look at the number in the ones place. If this number is less than 5, you must round the tens number down. If this number is 5 or more, you must round the tens number up.
8	C	Moving from right to left, the positions are as follows: ones, tens, hundreds. In order to round to the nearest hundred, you must look at the number in the tens place. If this number is less than 5, you must round the hundreds number down. If this number is 5 or more, you must round the hundreds number up.

Question No.	Answer	Detailed Explanation
9	B	Moving from right to left, the positions are as follows: ones, tens. In order to round to the nearest ten, you must look at the number in the ones place. If this number is less than 5, you must round the tens number down. If this number is 5 or more, you must round the tens number up.
10	C	Moving from right to left, the positions are as follows: ones, tens, hundreds, thousands. In order to round to the nearest hundred, you must look at the number in the tens place. If this number is less than 5, you must round the hundreds number down. If this number is 5 or more, you must round the hundreds number up. Option C is the only choice that would not round to 1,000. It would round to 1,100.
11	C	Moving from right to left, the positions are as follows: ones, tens, hundreds. In order to round to the nearest ten, you must look at the number in the ones place. If this number is less than 5, you must round the tens number down. If this number is 5 or more, you must round the tens number up. With these rules, there are 5 numbers that would round up to 100 (95, 96, 97, 98, and 99), there are 4 numbers that would round down to 100 (101,102, 103, and 104), and 100 rounds to itself. This is 10 numbers in all.
12	C	Moving from right to left, the positions are as follows: ones, tens, hundreds. In order to round to the nearest ten, you must look at the number in the ones place. If this number is less than 5, you must round the tens number down. If this number is 5 or more, you must round the tens number up. 795 has 5 in ones place. So, 795 round to 800, when rounded to nearest ten. In order to round to the nearest hundred, you must look at the number in the tens place. If this number is less than 5, you must round the hundreds number down. If this number is 5 or more, you must round the hundreds number up. 795 has a 9 in its hundreds place. So, 795 would round to 800, when rounded to nearest hundred. So, in both the cases, rounded to nearest ten or hundred, 795 would round to 800.

Question No.	Answer	Detailed Explanation
13	B	Moving from right to left, the positions are as follows: ones, tens, hundreds, thousands. In order to round to the nearest hundred, you must look at the number in the tens place. If this number is less than 5, you must round the hundreds number down. If this number is 5 or more, you must round the hundreds number up. 1,090 has a 9 in its tens place, so when rounding to the nearest hundred, it would round to 1,100. If 1,090 is rounded to nearest ten, it would have been the same. So, option (A) and (C) are wrong. If 1,090 is rounded to nearest thousand, it would have been 1,000. So, option (D) is also wrong.
14	B	Moving from right to left, the positions are as follows: ones, tens, hundreds, thousands. In order to round to the nearest ten, you must look at the number in the ones place. If this number is less than 5, you must round the tens number down. If this number is 5 or more, you must round the tens number up.
15	A.	Moving from right to left, the positions are as follows: ones, tens, hundreds. In order to round to the nearest hundred, you must look at the number in the tens place. If this number is less than 5, you must round the hundreds number down. If this number is 5 or more, you must round the hundreds number up. Option A is the only choice that would not fit this criteria to round to 800.
16	A & B	When rounding to the nearest hundred look at the number in the tens place. If the number is less than 5 round down to the nearest hundred. If the number is 5 or more, round up to the nearest hundred. 617 has 1 in tens place. So, 617 rounds down to 600, the nearest hundred. When rounding to the nearest ten look at the number in the ones place. If the number is less than 5 round down to the nearest ten. If the number is 5 or more, round up to the nearest ten. So, 617 rounds up to 620, the nearest ten.
17	500	When rounding to the nearest hundred look at the number in the tens place. If the number is less than 5 round down to the nearest hundred. If the number is 5 or more, round up to the nearest hundred. 489 is nearest to 500 on the number line.

Question No.	Answer	Detailed Explanation
18		

Number	Number when rounded to the nearest ten	Number when rounded to the nearest hundred
2,349	2,350	2,300
4,092	**4,090**	**4,100**
8,396	**8,400**	**8,400**

Chapter 2: Computation and Algebraic Thinking

Lesson 1: Addition & Subtraction

1. What is the standard form of 70,000 + 6,000 + 800 + 60 + 2?

   Ⓐ 706,862
   Ⓑ 76,862
   Ⓒ 7,682
   Ⓓ 782

2. Two numbers have a difference of 29. The two numbers could be _____.

   Ⓐ 11 and 18
   Ⓑ 23 and 42
   Ⓒ 40 and 11
   Ⓓ 50 and 39

3. Two numbers add up to 756. One number is 356. What is the other number?

   Ⓐ 356
   Ⓑ 300
   Ⓒ 400
   Ⓓ 456

4. Which of these expressions has the same difference as 94 - 50?

   Ⓐ 70 - 34
   Ⓑ 80 - 46
   Ⓒ 60 - 16
   Ⓓ 90 - 54

5. Which of these number sentences is not true?

   Ⓐ 88 + 12 = 90 + 10
   Ⓑ 82 + 18 = 88 + 12
   Ⓒ 56 + 45 = 54 + 56
   Ⓓ 46 + 15 = 56 + 5

6. Jim has 640 baseball cards and 280 basketball cards. How many sports cards does Jim have in all?

   Ⓐ 820 cards
   Ⓑ 360 cards
   Ⓒ 8,120 cards
   Ⓓ 920 cards

7. Find the difference.
   860 - 659

   Ⓐ 219
   Ⓑ 319
   Ⓒ 201
   Ⓓ 19

8. The students made 565 book covers for their math books. They used up 422 of the book covers. How many book covers are left?

   Ⓐ 242 book covers
   Ⓑ 163 book covers
   Ⓒ 987 book covers
   Ⓓ 143 book covers

9. What is the difference of 32 and 5?

   Ⓐ 33
   Ⓑ 27
   Ⓒ 160
   Ⓓ 37

10. Jenny plans to sell 50 boxes of cookies to help her scout troop raise funds. She sold 20 boxes to her neighbors. Her dad sold 15 boxes at his work office. How many more boxes does she need to sell to meet her goal?

    Ⓐ 15 boxes
    Ⓑ 10 boxes
    Ⓒ 5 boxes
    Ⓓ 25 boxes

11. Sara had 124 stickers. She gave away 62 stickers and bought 73 more stickers. How many stickers does Sara have now?

    Ⓐ 135 stickers
    Ⓑ 62 stickers
    Ⓒ 120 stickers
    Ⓓ 11 stickers

12. Which of these addition expressions would require regrouping of hundreds and tens?

    Ⓐ 923 + 37
    Ⓑ 456 + 443
    Ⓒ 235 + 234
    Ⓓ 576 + 442

13. There were 605 people sitting in an auditorium at the start of a show. Thirty-five people left during the intermission. How many people remained in the auditorium after the intermission?

    Ⓐ 630 people
    Ⓑ 580 people
    Ⓒ 570 people
    Ⓓ 595 people

14. If 3 tens are subtracted from 401, what is the difference?

    Ⓐ 471
    Ⓑ 371
    Ⓒ 398
    Ⓓ 381

15. Find the sum of 37 + 93 + 200.

    Ⓐ 330
    Ⓑ 663
    Ⓒ 320
    Ⓓ 300

16. Select the correct sum for each addition expression.

	899	467	558
422 + 136	○	○	○
608 + 291	○	○	○
157 + 310	○	○	○

17. Hannah received a score of 604 on the exam. Ben received a score of 719. What was the difference between the two scores? Write your answer in the box given below.

18. Type in the correct numbers to make the sum true.

	Hundreds	Tens	Ones
	2		5
+		3	
Total	8	4	9

19. Karen has 805 milliliters of milk. After he drinks some milk, 538 milliliters are left. How much milk did Karen drink? Show the steps by which you arrive at the answer.

CHAPTER 2 → Lesson 2: Two-Step Problems

1. Danny has 47 baseball cards. He gives his brother 11 cards. Danny then divides the remaining cards between 3 of his classmates. How many cards does each classmate receive?

    Ⓐ 15
    Ⓑ 3
    Ⓒ 12
    Ⓓ 11

2. Two classes of grade three students are lined up outside. One class is lined up in 3 rows of 7. The other class is lined up in 4 rows of 5. How many total third graders are lined up outside?

    Ⓐ 19 third graders
    Ⓑ 21 third graders
    Ⓒ 41 third graders
    Ⓓ 20 third graders

3. Jessica earns 10 dollars per hour for babysitting. She has saved 60 dollars so far. How many more hours will she need to babysit to buy something that costs 100 dollars?

    Ⓐ 40 hours
    Ⓑ 6 hours
    Ⓒ 10 hours
    Ⓓ 4 hours

4. George started with 2 bags of 10 cookies. He gave 12 cookies to his parents. How many cookies does George have now?

    Ⓐ 8 cookies
    Ⓑ 10 cookies
    Ⓒ 12 cookies
    Ⓓ 20 cookies

5. Renae has 60 minutes to do her chores and do her homework. She has 3 chores to complete and each chore takes 15 minutes to complete. After completing her chores, how many minutes does Renae have left to do her homework?

    Ⓐ 15 minutes
    Ⓑ 45 minutes
    Ⓒ 30 minutes
    Ⓓ 0 minutes

6. Anna and Jamie want to buy a new board game. The original cost was 28 dollars. It is on sale for 4 dollars off. How much money should each girl pay if they buy the game on sale and pay equal amounts?

   Ⓐ $24
   Ⓑ $2
   Ⓒ $12
   Ⓓ $14

7. 100 students went on a field trip. Ten students rode with their parents in a car while the remaining students were divided equally into 5 buses. How many students rode on each bus?

   Ⓐ 9 students
   Ⓑ 18 students
   Ⓒ 50 students
   Ⓓ 90 students

8. Julia has 32 books. Her sister has twice the number of books that Julia has. How many books do the girls have altogether?

   Ⓐ 66 books
   Ⓑ 32 books
   Ⓒ 64 books
   Ⓓ 96 books

9. Alicia bought 5 crates of apples. Each crate had 8 apples. She divided the apples equally into 10 bags. How many apples were in each bag?

   Ⓐ 40 apples
   Ⓑ 4 apples
   Ⓒ 10 apples
   Ⓓ 5 apples

10. Janeth went to the store and spent 4 dollars on markers. She also bought 3 copies of the same book. If she spent a total of 19 dollars, how much did each book cost?

    Ⓐ 5 dollars
    Ⓑ 4 dollars
    Ⓒ 3 dollars
    Ⓓ 6 dollars

11. Brian won 24 candy bars in a contest. He gave 2 candy bars to each of his 7 friends. How many candy bars does Brian have left?

    Ⓐ 14 candy bars
    Ⓑ 12 candy bars
    Ⓒ 10 candy bars
    Ⓓ 17 candy bars

12. Jenine gave 3 mini cupcakes to each of her three sisters. She then had 4 left for herself. How many mini cupcakes did Jenine start with?

    Ⓐ 13 mini cupcakes
    Ⓑ 9 mini cupcakes
    Ⓒ 10 mini cupcakes
    Ⓓ 7 mini cupcakes

13. Twenty-two people visited the art exhibit at the museum on Friday. Twice as many people visited on Saturday. How many people combined visited the art exhibit at the museum on Friday and Saturday?

    Ⓐ 88 people
    Ⓑ 66 people
    Ⓒ 22 people
    Ⓓ 44 people

14. Audrey can watch 5 hours of TV a week. She has already watched 4 shows that are each 1 hour long. How many more hours can she watch TV this week?

    Ⓐ 3 hours
    Ⓑ 2 hours
    Ⓒ 1 hour
    Ⓓ 4 hours

15. Greg had 3 books. His older brother gave him 15 more books. Greg wants to divide his total number of books equally onto 6 shelves. How many books should he place on each shelf?

    Ⓐ 3 books
    Ⓑ 12 books
    Ⓒ 18 books
    Ⓓ 6 books

16. Sarah bought 2 boxes of doughnuts. Each box contained 12 donuts. She shared a total of 7 donuts with her friends. How many doughnuts does she have now? Identify which equations can be used to find the answer. (Choose all correct answers)

    Ⓐ 12 x 2= 24
    Ⓑ 12 + 2= 14
    Ⓒ 24-7= 17
    Ⓓ 2 + 12 + 7= 21

17. Freddy has a collection of 32 baseball cards. He wants to share the cards with 4 classmates. One of the classmates brings 8 more cards to add to the collection. If Freddy and his classmates share all the cards, each receiving the same number, how many cards does each person have? What should be the steps to be followed to arrive at the answer? Write the steps in the correct sequence in the boxes given below.

    Ⓐ 32 - 6 = 26
    Ⓑ 4 + 1 = 5
    Ⓒ 32 + 8 = 40
    Ⓓ 40 ÷ 5 = 8

18. A farmer collected 22 pints of milk from his cows. He put all the milk into bottles. Each bottle holds 2 pints of milk. He accidentally spilled 6 bottles of milk. How many bottles are left with the farmer now? Circle the math sentences that can be used to find the answer. (Circle all correct answers)

    Ⓐ 11
    Ⓑ 24
    Ⓒ 16
    Ⓓ 5

19. For each of the problems in the first column, select the correct answer.

	$50	$5	$10	$4
Karen had 86 dollars. He bought 7 books. After buying them he had 16 dollars. How much did each book cost?	○	○	○	○
Jose and his four friends bought a new board game. It was on sale for 20 dollars off. If each of the boys (total 5 of them) paid $6. What was the original cost of the new board game?	○	○	○	○
A shopkeeper buys 5 pens for $35 and sells them at the rate of $8 per pen. If he sells all the five pens, how much profit he will get?	○	○	○	○
Jeffrey bought 8 actions figures which cost 3 dollars each from John. John bought 6 books from the amount he received from Jeffrey. If the cost of each book John purchased is the same, what is the cost of each book?	○	○	○	○

CHAPTER 2 → Lesson 3: Understanding Multiplication

1. Which multiplication fact is being modeled below?

   Ⓐ 3 x 10 = 30
   Ⓑ 4 x 10 = 40
   Ⓒ 4 x 9 = 36
   Ⓓ 3 x 9 = 27

2. Which numerical expression describes this array?

   Ⓐ 4 + 5
   Ⓑ 5 + 4
   Ⓒ 4 x 5
   Ⓓ 4 x 4

3. Which number sentence describes this array?

   Ⓐ 8 x 4 = 32
   Ⓑ 7 + 5 = 12
   Ⓒ 5 x 7 = 35
   Ⓓ 4 x 7 = 28

4. Which number sentence describes this array?

　○○○○○○○○○○○○
　○○○○○○○○○○○○

Ⓐ 2 x 12 = 24
Ⓑ 2 + 12 = 14
Ⓒ 12 + 2 = 24
Ⓓ 10 x 2 = 20

5. Identify the multiplication sentence for the picture below:

Ⓐ 4 x 4 = 16
Ⓑ 4 x 3 = 12
Ⓒ 3 x 4 = 12
Ⓓ 4 x 2 = 8

6. What multiplication fact does this picture model?

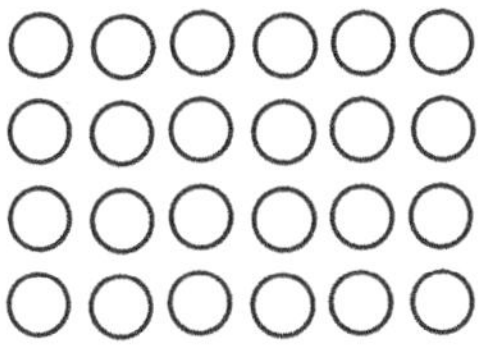

　○○○○○○
　○○○○○○
　○○○○○○
　○○○○○○

Ⓐ 4 x 6 = 24
Ⓑ 4 x 7 = 28
Ⓒ 6 x 3 = 18
Ⓓ 7 x 4 = 28

7. Identify the multiplication sentence for the picture below:

☺ ☺ ☺ ☺ ☺ ☺
☺ ☺ ☺ ☺ ☺ ☺
☺ ☺ ☺ ☺ ☺ ☺

Ⓐ 7 x 2 = 14
Ⓑ 7 x 3 = 21
Ⓒ 7 x 4 = 28
Ⓓ 6 x 3 = 18

8. Identify the multiplication sentence for the picture below:

Ⓐ 4 x 4 = 16
Ⓑ 3 x 6 = 18
Ⓒ 3 x 4 = 12
Ⓓ 3 x 5 = 15

9. Identify the multiplication sentence for the picture below:

Ⓐ 3 x 2 = 6
Ⓑ 3 x 3 = 9
Ⓒ 4 x 2 = 8
Ⓓ 3 x 1 = 3

10. Identify the multiplication sentence for the picture below:

Ⓐ 3 x 5 = 15
Ⓑ 4 x 4 = 16
Ⓒ 5 x 4 = 20
Ⓓ 7 x 4 = 28

11. Identify the multiplication sentence for the picture below:

Ⓐ 2 x 5 = 10
Ⓑ 4 x 2 = 8
Ⓒ 4 x 1 = 4
Ⓓ 4 + 2 = 6

12. Identify the multiplication sentence for the picture below:

Ⓐ 6 x 7 = 42
Ⓑ 6 x 8 = 48
Ⓒ 8 x 9 = 72
Ⓓ 8 x 8 = 64

13. Identify the multiplication sentence for the picture below:

Ⓐ 10 x 1 = 10
Ⓑ 9 x 2 = 18
Ⓒ 2 x 10 = 20
Ⓓ 5 x 4 = 20

14. Identify the multiplication sentence for the picture below:

Ⓐ 5 x 5 = 25
Ⓑ 4 x 4 = 16
Ⓒ 4 x 6 = 24
Ⓓ 5 x 4 = 20

15. Identify the multiplication sentence for the picture below

Ⓐ 3 x 1 = 3
Ⓑ 5 x 3 = 15
Ⓒ 3 x 2 = 6
Ⓓ 3 x 3 = 9

16. Represent the below equation as a multiplication expression. Write your answer in the box below.

    8 + 8 + 8 + 8?

    ☐

17. Match each multiplication statement to the correct addition statement by darkening the corresponding circles.

	Column A: 3+3+3+3+3+3	Column B: 3+3+3+3+3+3+3+3	Column C: 3+3+3
3 x 8	○	○	○
3 x 3	○	○	○
3 x 6	○	○	○

18. For each of the pictures, write the correct mathematical expression in the box.

☐

☐

☐

19. John finds the solution for 8 x 6 by solving for (8 x 5) + 8. Is John correct? Explain why you think that John's strategy is correct or not? Write your answer in the box below.

20. There are Seven boys, and each of them buys 6 pens. How many pens do they buy all together? Write an equation to represent this. Also, Find the total number of pens purchased using the equation.

21. Complete the following table:

Number of lions	5	6	9		
Total number of legs	20			32	16

CHAPTER 2 → Lesson 4: Understanding Division

1. George is canning pears. He has 100 pears and he divides the pears evenly among 10 pots. How many pears does George put in each pot?

   Ⓐ 9 pears
   Ⓑ 5 pears
   Ⓒ 8 pears
   Ⓓ 10 pears

2. Marisa made 15 woolen dolls. She gave the same number of woolen dolls to 3 friends. How many dolls did Marisa give to each friend?

   Ⓐ 4 woolen dolls
   Ⓑ 3 woolen dolls
   Ⓒ 5 woolen dolls
   Ⓓ 6 woolen dolls

3. Lisa bought 50 mangoes. She divided them equally into 5 basins. How many mangoes did Lisa put in each basin?

   Ⓐ 10 mangos
   Ⓑ 8 mangos
   Ⓒ 5 mangos
   Ⓓ 7 mangos

4. Jennifer picked 30 oranges from the basket. If it takes 6 oranges to make a one liter jar of juice, how many one liter jars of juice can Jennifer make?

   Ⓐ 4 jars
   Ⓑ 3 jars
   Ⓒ 6 jars
   Ⓓ 5 jars

5. Miller bought 80 rolls of paper towels. If there are 10 rolls of paper towels in each pack, how many packs of paper towels did Miller buy?

   Ⓐ 6 packs
   Ⓑ 8 packs
   Ⓒ 7 packs
   Ⓓ 5 packs

6. James takes 15 photographs of his school building. He gave the same number of photographs to 5 friends. How many photographs did James give to each friend?

   Ⓐ 2 photographs
   Ⓑ 3 photographs
   Ⓒ 6 photographs
   Ⓓ 5 photographs

7. Ron took 81 playing cards and arranged them into 9 equal piles. How many playing cards did Ron put in each pile?

   Ⓐ 5 playing cards
   Ⓑ 4 playing cards
   Ⓒ 6 playing cards
   Ⓓ 9 playing cards

8. Robert wants to buy 40 ice cream cups from the ice cream parlor. If there are 10 ice cream cups in each box, how many boxes of ice cream cups should Robert buy?

   Ⓐ 6 boxes
   Ⓑ 3 boxes
   Ⓒ 4 boxes
   Ⓓ 5 boxes

9. Marilyn wants to purchase 20 tiles. If the tiles come in packs of 5, how many packs should Marilyn buy?

   Ⓐ 3 packs
   Ⓑ 4 packs
   Ⓒ 5 packs
   Ⓓ 6 packs

10. There are 30 people running around the path. If the runners are evenly divided among the path's 5 lanes, how many people are running in each lane?

    Ⓐ 6 runners
    Ⓑ 5 runners
    Ⓒ 8 runners
    Ⓓ 4 runners

LumosLearning.com

11. Sally is buying goodie bags for her class. She needs 24 bags in all. If the bags come in packs of 3, how many packs does Sally need?

    Ⓐ 21 packs
    Ⓑ 3 packs
    Ⓒ 24 packs
    Ⓓ 8 packs

12. Mr. Johnson is planting a garden. He wants to use all of his 44 seeds and wants to make 4 rows of vegetables. How many seeds should he plant in each row?

    Ⓐ 22 seeds
    Ⓑ 11 seeds
    Ⓒ 4 seeds
    Ⓓ 88 seeds

13. Destiny, Jimmy, and Marcy have 32 marbles all together. Tommy adds 4 marbles to the set. If the group of friends wants to evenly divide the marbles so that each person has the same number, how many marbles should each person receive?

    Ⓐ 4 marbles
    Ⓑ 8 marbles
    Ⓒ 9 marbles
    Ⓓ 10 marbles

14. Mr. Baker earned $100 for five days of work. If he made the same amount each day, how much money did he make per day?

    Ⓐ $15 per day
    Ⓑ $20 per day
    Ⓒ $25 per day
    Ⓓ $30 per day

15. Seth and his brother have collected 26 seashells on the beach. If they want to share them equally, how many seashells will each of them receive?

    Ⓐ 6 seashells
    Ⓑ 9 seashells
    Ⓒ 26 seashells
    Ⓓ 13 seashells

16. A pizza is cut into 8 slices. Tim and Kira want to share the pizza. If they both eat the same number of slices, how many slices will each person eat? Write it in the box given below.

17. Gabriela has 16 stickers. She wants to find two ways to divide the stickers into equal groups. Which expressions can she use to divide the stickers? Mark all the correct answers.

   Ⓐ 16 ÷ 2
   Ⓑ 16 ÷ 3
   Ⓒ 16 ÷ 4
   Ⓓ 16 ÷ 5

18. Circle the picture that shows the expression 10 ÷ 5.

   Ⓐ

   Ⓑ

   Ⓒ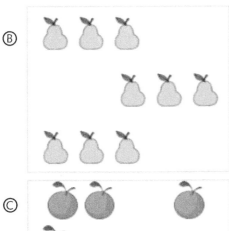

19. Miriam has 100 marbles. She wants to divide the marbles into equal groups. How many ways can she do this? Complete the table by listing all the possible ways in which you can divide 100. Write the missing numbers and fill the table. Enter the numbers in ascending order.

Dividend	Possible Divisors
100	2
100	4
100	
100	
100	20
100	25
100	

20. For each expression below, choose the correct symbol to be filled in the blank.

	<	>	=
30 ÷ 5 ___ 42 ÷ 6	○	○	○
72 ÷ 8 ___ 63 ÷ 7	○	○	○
54 ÷ 6 ___ 56 ÷ 7	○	○	○

CHAPTER 2 → Lesson 5: Multiplication & Division Facts

1. Find the product.
   6 x 0 = ___

   Ⓐ 6
   Ⓑ 1
   Ⓒ 0
   Ⓓ 2

2. Find the product.
   1 x 10 = ___

   Ⓐ 0
   Ⓑ 1
   Ⓒ 10
   Ⓓ 11

3. Solve.
   3 x 8 = ___

   Ⓐ 24
   Ⓑ 21
   Ⓒ 18
   Ⓓ 28

4. Solve.
   ___ = 5 x 9

   Ⓐ 40
   Ⓑ 45
   Ⓒ 50
   Ⓓ 35

5. Find the product of 8 and 6.

   Ⓐ 14
   Ⓑ 42
   Ⓒ 48
   Ⓓ 56

6. Find the product of 7 and 7.

    Ⓐ 42
    Ⓑ 46
    Ⓒ 49
    Ⓓ 56

7. Find the product of 4 and 6.

    Ⓐ 20
    Ⓑ 24
    Ⓒ 28
    Ⓓ 32

8. Find the product.
   6 x 9 = ___

    Ⓐ 54
    Ⓑ 45
    Ⓒ 48
    Ⓓ 64

9. Find the product.
   ___ = 9 x 8

    Ⓐ 64
    Ⓑ 72
    Ⓒ 81
    Ⓓ 82

10. Which expression below has a product of 48?

    Ⓐ 6 x 7
    Ⓑ 4 x 14
    Ⓒ 7 x 8
    Ⓓ 8 x 6

11. Find the quotient of 25 and 5.

    Ⓐ 20
    Ⓑ 5
    Ⓒ 4
    Ⓓ 15

12. What is 32 divided by 4?

    Ⓐ 9
    Ⓑ 8
    Ⓒ 7
    Ⓓ 6

13. What is 28 divided by 7?

    Ⓐ 4
    Ⓑ 5
    Ⓒ 3
    Ⓓ 6

14. Find the quotient.
    0 ÷ 5 = ___

    Ⓐ 0
    Ⓑ 1
    Ⓒ 5
    Ⓓ 50

15. Find the quotient.
    7 ÷ 1 = ___

    Ⓐ 1
    Ⓑ 0
    Ⓒ 7
    Ⓓ 8

16. Find the quotient.
    ___ = 12 ÷ 2

    Ⓐ 9
    Ⓑ 8
    Ⓒ 7
    Ⓓ 6

17. Divide.
    63 ÷ 9 = ___

    Ⓐ 6
    Ⓑ 7
    Ⓒ 8
    Ⓓ 9

LumosLearning.com

18. Divide.
    42 ÷ 7 = ___

    Ⓐ 5
    Ⓑ 6
    Ⓒ 7
    Ⓓ 8

19. Find the quotient of 33 and 3.

    Ⓐ 11
    Ⓑ 12
    Ⓒ 10
    Ⓓ 9

20. Divide.
    56 ÷ 7 = ___

    Ⓐ 6
    Ⓑ 7
    Ⓒ 8
    Ⓓ 9

21. Solve.
    4 x 12 = ___

    Ⓐ 36
    Ⓑ 48
    Ⓒ 42
    Ⓓ 46

22. Solve.
    ___ = 75 ÷ 5

    Ⓐ 13
    Ⓑ 15
    Ⓒ 17
    Ⓓ 25

23. Solve.
    84 ÷ 12 = ___

    Ⓐ 7
    Ⓑ 8
    Ⓒ 9
    Ⓓ 12

24. Solve.
    12 x 3 = ____
    - Ⓐ 32
    - Ⓑ 36
    - Ⓒ 39
    - Ⓓ 48

25. Solve.
    36 ÷ 3 = ____
    - Ⓐ 22
    - Ⓑ 12
    - Ⓒ 14
    - Ⓓ 18

26. Solve.
    60 ÷ 5 = ____
    - Ⓐ 8
    - Ⓑ 9
    - Ⓒ 12
    - Ⓓ 14

27. Solve.
    11 x 4 = ____
    - Ⓐ 32
    - Ⓑ 44
    - Ⓒ 39
    - Ⓓ 46

28. Solve.
    ____ = 80 ÷ 8
    - Ⓐ 10
    - Ⓑ 9
    - Ⓒ 8
    - Ⓓ 12

29. Solve.

12 x 8 = _____

Ⓐ 72
Ⓑ 84
Ⓒ 92
Ⓓ 96

30. Solve.

50 ÷ 5 = _____

Ⓐ 25
Ⓑ 20
Ⓒ 10
Ⓓ 50

31. Match each equation to the correct product by shading the appropriate circle.

	12	18	32
6 x 3=	○	○	○
8 x 4=	○	○	○
4 x 3=	○	○	○
9 x 2=	○	○	○

32. Solve and write the answer in the box given below.

24 ÷ 4=?

33. Which expressions have a product of 36. Select all the correct answers.

Ⓐ 6 x 6
Ⓑ 8 x 4
Ⓒ 6 x 8
Ⓓ 9 x 4

34. Find the quotient.

    27 ÷ 9

    Ⓐ 18
    Ⓑ 4
    Ⓒ 3
    Ⓓ 2

35. Complete the following table.

5	x	8	=	
8	÷		=	8
	÷	7	=	0
6	x		=	30

CHAPTER 2 → Lesson 6: Multiplication & Division Properties

1. Which of these statements is not true?

    Ⓐ 4 x (3 x 6) = (4 x 3) x 6
    Ⓑ 4 x 3 = 3 x 4
    Ⓒ 15 x 0 = 0 x 15
    Ⓓ 12 x 1 = 12 x 12

2. Which of these statements is true?

    Ⓐ The product of 11 x 6 is equal to the product of 6 x 11.
    Ⓑ The product of 11 x 6 is greater than the product of 6 x 11.
    Ⓒ The product of 11 x 6 is less than the product of 6 x 11.
    Ⓓ There is no relationship between the product of 11 x 6 and the product of 6 x 11.

3. Which of the following expressions has a value of 0?

    Ⓐ (3 x 4) x 1
    Ⓑ 50 x 1
    Ⓒ 3 x 4 x 0
    Ⓓ (3 x 1) x 2

4. Which set of two numerical expressions have a value of zero?

    Ⓐ 60 x 1 and 1 x 60
    Ⓑ 10 x 10 and 0 x 10
    Ⓒ 27 x 0 and 0 x 27
    Ⓓ 0 ÷ 15 and 15 ÷ 15

5. Which mathematical property does this equation model?
   6 x 1 = 6

    Ⓐ Commutative Property of Multiplication
    Ⓑ Associative Property of Multiplication
    Ⓒ Identity Property of Multiplication
    Ⓓ Distributive Property

6. Which mathematical property does this equation model?
   9 x 6 = 6 x 9

   Ⓐ Commutative Property of Multiplication
   Ⓑ Associative Property of Multiplication
   Ⓒ Identity Property of Multiplication
   Ⓓ Distributive Property

7. Which mathematical property does this equation model?
   (2 x 10) x 3 = 2 x (10 x 3)

   Ⓐ Commutative Property of Multiplication
   Ⓑ Associative Property of Multiplication
   Ⓒ Identity Property of Multiplication
   Ⓓ Distributive Property

8. Which mathematical property does this equation model?
   4 x (9 + 6) = (4 x 9) + (4 x 6)

   Ⓐ Commutative Property of Multiplication
   Ⓑ Associative Property of Multiplication
   Ⓒ Identity Property of Multiplication
   Ⓓ Distributive Property

9. By the Commutative Property of Multiplication, if you know that 4 x 5= 20, then you also know that _____ .

   Ⓐ 20 is an even number
   Ⓑ 4 x 6 = 24
   Ⓒ 5 x 4 = 20
   Ⓓ 5 is greater than 4

10. By the Associative Property of Multiplication, If you know that (2 x 3) x 4 = 24, then you also know that _____.

    Ⓐ 2 x (3 x 4) = 24
    Ⓑ 2 x 4 = 8
    Ⓒ 24 ÷ 6 = 4
    Ⓓ (2 x 3) x 5 = 30

11. Complete the following statement:
    Multiplication and _____ are inverse operations.

    Ⓐ addition
    Ⓑ subtraction
    Ⓒ division
    Ⓓ distribution

12. Below equation models the _____.
    32 x 7 = 7 x 32

    Ⓐ Commutative Property of Multiplication
    Ⓑ Associative Property of Multiplication
    Ⓒ Identity Property of Multiplication
    Ⓓ Distributive Property

13. Below equation models the _____.
    26 x 2 = (20 x 2) + (6 x 2)

    Ⓐ Commutative Property of Multiplication
    Ⓑ Associative Property of Multiplication
    Ⓒ Identity Property of Multiplication
    Ⓓ Distributive Property

14. By the Identity Property of Multiplication, you know that _____.

    Ⓐ 2 x 2 = 4
    Ⓑ 0 x 0 = 0
    Ⓒ 6 x 1 = 6
    Ⓓ 5 ÷ 5 = 1

15. What number belongs in the blank?
    10 x __ = 10

    Ⓐ 1
    Ⓑ 0
    Ⓒ 10
    Ⓓ 5

16. From the below 4 options, select 2 options that will result in the same product based on the commutative property of multiplication?

    Ⓐ 5 + 4
    Ⓑ 10 x 2
    Ⓒ 5 x 4
    Ⓓ 4 x 5

17. Make the equation true according to the Identity Property of Multiplication. Write the correct number in the answer box given below.

    7 x ___ = 7

18. (2 x 3) x 4 = 24 and 2 x (3 x 4) = 24

    Identify the property that is applicable. Circle the correct answer choice.

    Ⓐ Associative property of multiplication
    Ⓑ Distributive property
    Ⓒ Commutative property of multiplication
    Ⓓ Identity Property of Multiplication

19. **PART A:**

Which number will make the below equation true. Write your answer in the box given below.

3 x 7 = 7 x ?

☐

PART B:

Which property did you use in Part A to arrive at the answer. Write the name of the property in the box given below.

☐

20. Match the property with the correct example by shading the appropriate circle.

	3 x (5 x 7) = (3 x 5) x 7	3 x 1 = 3	3 x 5 = 5 x 3	3 x (5 + 7) = (3 x 5) + (3 x 7)
Commutative Property	○	○	○	○
Associative Property	○	○	○	○
Identity Property	○	○	○	○
Distributive Property	○	○	○	○

CHAPTER 2 → Lesson 7: Applying Multiplication & Division

1. The product in this number sentence is _____.
   54 x 3 = ?

   Ⓐ 54
   Ⓑ 162
   Ⓒ 3
   Ⓓ 54 and 3

2. A Snack Shop has twice as many popcorn balls as they do cotton candy. If there are 30 popcorn balls, how many cotton candies are there?

   Ⓐ 7
   Ⓑ 450
   Ⓒ 30
   Ⓓ 15

3. Monica has 56 DVDs in her movie collection. This is 8 times as many as Sue has. How many DVDs does Sue have?

   Ⓐ 8
   Ⓑ 6
   Ⓒ 7
   Ⓓ 10

4. Jonathan can do 7 jumping jacks. Marcus can do 4 times as many as Jonathan. How many jumping jacks can Marcus do?

   Ⓐ 28
   Ⓑ 8
   Ⓒ 4
   Ⓓ 7

5. Darren has seen 4 movies this year. Marsha has seen 3 times as many movies as Darren. How many movies has Marsha seen?

   Ⓐ 7
   Ⓑ 3
   Ⓒ 4
   Ⓓ 12

6. Sarah is planting a garden. She will plant 4 rows with 9 seeds in each row. How many plants will be in the garden?

   Ⓐ 32 plants
   Ⓑ 36 plants
   Ⓒ 42 plants
   Ⓓ 13 plants

7. Mrs. Huerta's class is having a pizza party. There are 24 students in the class. Each pizza has 12 slices. How many pizzas does Mrs. Huerta need to order for each child to have 1 slice?

   Ⓐ 3 pizzas
   Ⓑ 2 pizzas
   Ⓒ 1 pizza
   Ⓓ 4 pizzas

8. There are 27 apples. How many pies can be made if each pie uses 3 apples?

   Ⓐ 7 pies
   Ⓑ 8 pies
   Ⓒ 9 pies
   Ⓓ 10 pies

9. Keegan is planting a garden in even rows. He has 48 seeds. Which layout is NOT possible?

   Ⓐ 6 rows of 8 seeds
   Ⓑ 8 rows of 6 seeds
   Ⓒ 7 rows of 7 seeds
   Ⓓ 12 rows of 4 seeds

10. There are 25 students in a gym class. They want to play a game with 5 equal teams. How many students will be on each team?

    Ⓐ 4 students
    Ⓑ 5 students
    Ⓒ 7 students
    Ⓓ 3 students

11. Josie has 7 days to read a book with 21 chapters. How many chapters should she read each day?

   Ⓐ 3 chapters
   Ⓑ 4 chapters
   Ⓒ 5 chapters
   Ⓓ 7 chapters

12. Devon has $40 to spend on fuel. One gallon of fuel costs $5. How many gallons can Devon afford to buy?

   Ⓐ 5 gallons
   Ⓑ 12 gallons
   Ⓒ 9 gallons
   Ⓓ 8 gallons

13. Amanda is using the following cake recipe:
    4 cups flour
    1 cup sugar
    3 cups milk
    1 egg
    If Amanda needs to make three batches, how many cups of flour will she need?

   Ⓐ 7 cups
   Ⓑ 12 cups
   Ⓒ 10 cups
   Ⓓ 16 cups

14. Kim invited 20 friends to her birthday party. Twice as many friends than she invited showed up the day of the party. Which number sentence could be used to solve how many friends came to the party?

   Ⓐ n + 20 = 2
   Ⓑ n x 20 = 2
   Ⓒ 20 x 2 = n
   Ⓓ 20 - n = 20

15. The product of 9 and a number is 45.
    Which number sentence models this situation?

   Ⓐ 9 + n = 45
   Ⓑ 45 + 9 = n
   Ⓒ 9 x n = 45
   Ⓓ 5 x n = 45

16. A classroom has 5 rows of desks. There are 6 desks in each row. How many desks are there altogether? Select the number sentences that represent the solution. Choose all correct answers.

    Ⓐ 6 - 5= 1
    Ⓑ 5 x 6= 30
    Ⓒ 6 x 5= 30
    Ⓓ 5 + 6= 11

17. Jasmine bought a bouquet of 24 flowers. She plans to give the same number of flowers to her 4 friends, Daniel, Raquel, Elliot and Sue. How many flowers will each friend receive? Circle the correct answer.

    Ⓐ 2
    Ⓑ 5
    Ⓒ 6
    Ⓓ 4

18. There are 48 cupcakes to be shared equally among 6 boys. How many cupcakes will each boy get? Write your answer in the box given below.

19. Fill in the blank with the correct operational symbol to make these equations true.

    A) $64 \div 8 = 2$ ___ 4.

    B) 2 ___ $4 = 42 \div 7$.

20. Joseph reads 8 pages every day. In how many days will he be able to complete reading a book which has 56 pages? Write an equation to represent this in the box below, and find the number of days Joseph takes to complete the book.

CHAPTER 2 → Lesson 8: Number Patterns

1. Which of the following is an even number?

   Ⓐ 764,723
   Ⓑ 90,835
   Ⓒ 5,862
   Ⓓ 609

2. Which of these sets contains no odd numbers?

   Ⓐ 13, 15, 81, 109, 199
   Ⓑ 123, 133, 421, 412, 600
   Ⓒ 34, 46, 48, 106, 88
   Ⓓ 12, 37, 6, 14, 144

3. Complete the following statement.
   The sum of two even numbers will always be _____ .

   Ⓐ greater than 10
   Ⓑ less than 100
   Ⓒ even
   Ⓓ odd

4. Complete the following statement.
   The product of two even numbers will always be _____ .

   Ⓐ even
   Ⓑ odd
   Ⓒ a multiple of 10
   Ⓓ a square number

5. Complete the following statement.
   A number has a nine in its ones place. The number must be a multiple of ____.

   Ⓐ 9
   Ⓑ 3
   Ⓒ 7
   Ⓓ None of the above

6. Complete the following statement.
   Numbers that are multiples of 8 are all _____.

   Ⓐ even
   Ⓑ multiples of 2
   Ⓒ multiples of 4
   Ⓓ All of the above

7. If this pattern continues, what will the next 3 numbers be?
   7, 14, 21, 28, 35, ...

   Ⓐ 41, 47, 53
   Ⓑ 49, 56, 63
   Ⓒ 77, 84, 91
   Ⓓ 42, 49, 56

8. Complete the following statement.
   All multiples of ____ can be decomposed into two equal addends.

   Ⓐ 6
   Ⓑ 9
   Ⓒ 3
   Ⓓ 5

9. If this pattern continues, what will the next 3 numbers be?
   9, 18, 27, 36, ...

   Ⓐ 54, 63, 72
   Ⓑ 45, 54, 63
   Ⓒ 44, 52, 60
   Ⓓ 44, 53, 62

10. A number is multiplied by 7 and the resulting product is even. Which of these could have been the number?

    Ⓐ 7
    Ⓑ 17
    Ⓒ 34
    Ⓓ 99

11. Complete the following statement.
    The multiples of 4 will always _____.

    Ⓐ have a 2 in the ones place
    Ⓑ be even
    Ⓒ be divisible by 8
    Ⓓ None of these

12. Complete the following statement.
    The sum of an even number and an odd number will always be _____.

    Ⓐ even
    Ⓑ odd
    Ⓒ divisible by 3
    Ⓓ None of the above

13. Complete the following statement.
    A multiple of 4 can have a _____ in its ones place.

    Ⓐ 2
    Ⓑ 8
    Ⓒ 6
    Ⓓ All of the above

14. Complete the following statement.
    A multiple of 5 can have a _____ as its ones digit.

    Ⓐ 0
    Ⓑ 3
    Ⓒ 9
    Ⓓ All of the above

15. Which of the following would produce an even product?

    Ⓐ an even number times an even number
    Ⓑ an even number times an odd number
    Ⓒ an odd number times an even number
    Ⓓ All of the above

16. Select the number that will come next if the pattern continues.

	10	35	24
2, 4, 6, 8	○	○	○
40, 36, 32, 28	○	○	○
7, 14, 21, 28	○	○	○

17. Type in the numbers that complete the table if the pattern is multiples of 3.

IN	OUT
3	9
4	
5	15
	18
7	

18. If the pattern continues, which of the following numbers will appear?
    Note: More than one option may be correct.

    100, 90, 80, 70

    Ⓐ 50
    Ⓑ 110
    Ⓒ 80
    Ⓓ 60

19. Complete the following statement. If you subtract an odd number from an even number, the difference will always be (a/an) _____. Circle the correct answer.

    Ⓐ Multiple of 3
    Ⓑ Even number
    Ⓒ Odd number
    Ⓓ Odd number or Even number

20. For each statement in the first column, choose all the correct answers.

	2	4	5	7
A number has a four in its ones place. The number can be a multiple of _____.	○	○	○	○
A number has a five in its ones place. The number can be a multiple of _____.	○	○	○	○
A number has a zero in its ones place. The number can be a multiple of _____.	○	○	○	○
A number has a three in its ones place. The number can be a multiple of _____.	○	○	○	○

End of Computation and Algebraic Thinking

ANSWER KEY AND DETAILED EXPLANATION

Chapter 2: Computation and Algebraic Thinking

Lesson 1: Addition & Subtraction

Question No.	Answer	Detailed Explanation
1	B	When these numbers are lined up correctly and then added, the results are: $$\begin{array}{r} 70000 \\ 6000 \\ 800 \\ 60 \\ +2 \\ \hline 76,862 \end{array}$$
2	C	Difference refers to the answer when two numbers are subtracted. Option C is the only choice where the difference will be 29.
3	C	In order to solve for the unknown number in this problem, subtract the two known numbers. 756 - 356 = 400.
4	C	The difference between 94 and 50 is 44. Option C is the only choice where 44 is also the answer.
5	C	In all of the choices, the sum on both sides of the number sentence must be equal. Option C is the only choice where the two sums are not equal. 56+45 = 101 and 54 + 56 = 110.
6	D	The phrase "in all" indicates that the two numbers must be added in order to find the total. 640 + 280 = 920.
7	C	$$\begin{array}{r} 860 \\ -659 \\ \hline 201 \end{array}$$
8	D	This problem involves subtracting the number of book covers used from the total number of book covers. $$\begin{array}{r} 565 \\ -422 \\ \hline 143 \end{array}$$

Question No.	Answer	Detailed Explanation
9	B	Difference refers to the answer when two numbers are subtracted. 32 - 5 ――― 27
10	A	First, add the number of cookies boxes that were sold. 20 + 15 = 35. Then subtract this number from the amount Jenny plans to sell to calculate the number still needed to sell. 50 - 35 = 15.
11	A	First, subtract 62 from 124 to see how many stickers Sara had after giving some away. 124 - 62 = 62. Then add 62 and 73 because she bought 73 more. This will calculate the total number of stickers. 62 + 73 = 135.
12	D	Option D is the only choice that requires regrouping of the hundreds and tens place. When the tens are added, 7 + 4 = 11. One is brought down while the ten is regrouped with the hundreds.
13	C	To calculate the number of people left in the auditorium, subtract the number of people that left from the total number of people. 605 - 35 ――― 570
14	B	3 tens is equal to 10 + 10 + 10 = 30. 401 - 30 ――― 371
15	A	2 0 0 9 3 + 3 7 ――――― 3 3 0
16		

	899	467	558
422 + 136	○	○	●
608 + 291	●	○	○
157 + 310	○	●	○

When adding three-digit numbers, start by adding the numbers in the ones place. Then add the numbers in the tens place. The last step is to add the numbers in the hundreds place.
422 + 136 = 558; 608 + 291 = 899; 157 + 310 = 467.

Question No.	Answer	Detailed Explanation
17	115	The correct answer is 115. The word 'difference' refers to the subtraction of numbers. When subtracting three-digit numbers, start by subtracting the numbers in the ones place. Then subtract the numbers in the tens place. The last step is to subtract the numbers in the hundreds place. 719 - 604 = 115.
18		

	Hundreds	Tens	Ones
	2	1	5
+	6	3	4
Total	8	4	9

The correct numbers are 6 (hundreds), 1 (tens), and 4 (ones). When adding three-digit numbers, start by adding the numbers in the ones place. Then add the numbers in the tens place. The last step is to add the numbers in the hundreds place. 215 + 634= 849.

| 19 | | This is a problem on subtraction. We have to subtract 538 from 805 to find the amount of milk Karen drank.

First, subtract the digits in the ones place. Since, 5 < 8, we try to borrow 1 from the tens place in the number 805. Since there is a zero in the tens place, we have to borrow 1 from the hundreds place and then borrow 1 from the tens place. (805 = 8 hundreds + 0 tens + 5 ones; it is rewritten as 7 hundreds + 10 tens + 5 ones = 7 hundreds + 9 tens + 15 ones); 15 - 8 = 7. So, in the answer, we have 7 in the ones place.

Next, subtract the digits in the tens place. 9 - 3 = 6. So, in the answer, we have 6 in the tens place.

Lastly, we subtract the digits in the hundreds place; 7 - 5 = 2. So, in the answer, we have 2 in the hundreds place.

805 - 538 = 267 |

Lesson 2: Two-Step Problems

Question No.	Answer	Detailed Explanation
1	C	First, calculate how many cards Danny has by subtracting 11 from 47; 47 - 11 = 36. Then divide this number by 3 to see how many each classmate will receive; 36 ÷ 3 = 12.
2	C	First, calculate how many students are in each class. The first class has 3 rows of 7 students and 3 x 7 = 21. The second class has 4 rows of 5 students and 4 x 5 = 20. Then add both totals to calculate the total number of students outside; 21 + 20 = 41.
3	D	First, calculate how much more money Jessica needs to save by subtracting what she has from what she needs; $100 - $60 = $40. Jessica needs 40 more dollars. Now divide 40 dollars by the amount she makes each hour of babysitting to find how many more hours she needs to work to earn the rest of the money; $40 ÷ 10 = 4.
4	A	First calculate how many cookies George began with by multiplying 2 and 10; 2 x 10 = 20. Then subtract the number of cookies he gave to his parents from this total; 20 - 12 = 8.
5	A	First calculate the total number of minutes Renae spends doing chores by multiplying 3 and 15; 3 x 15 = 45. Then subtract this number from 60 minutes to see how many minutes she has remaining to do her homework; 60 - 45 = 15.
6	C	First calculate the sale price of the game; $28 - 4 = $24. Then divide this answer by 2 to see how much each girl will pay; $24 ÷ 2 = $12.
7	B	First, calculate how many students rode the bus by subtracting the number of students who rode in cars from the total number of students; 100 - 10 = 90. Then divide this answer by the number of buses to see how many students rode on each bus; 90 ÷ 5 = 18.
8	D	First, calculate the number of books the sister has by multiplying the number of books Julia has by 2; 32 x 2 = 64. Then add this number to Julia's amount to get the total books; 64 + 32 = 96.
9	B	First, calculate the total number of apples Alicia has by multiplying 5 and 8; 5 x 8 = 40. Then calculate the number of apples in each bag by dividing the total number of apples by the number of bags; 40 ÷ 10 = 4.
10	A	First, subtract the cost of the markers from the total amount spent; $19 - $4 = $15. Then divide this answer by the number of books bought to calculate the cost of each book; $15 ÷ $3 = $5.

Question No.	Answer	Detailed Explanation
11	C	First, calculate how many candy bars were given to friends by multiplying the total number of friends by the number of bars each friend received; 7 x 2 = 14. Then subtract this answer from the total number of candy bars Brian won; 24 - 14 = 10.
12	A	First, calculate how many mini cupcakes were given to the sisters by multiplying the total number of sisters by the number each one received; 3 x 3 = 9. Then add this number to the number of cupcakes Jenine had left for herself; 9 + 4 = 13.
13	B	First, calculate how many people visited the museum on Saturday by multiplying the number of Friday visitors by 2 (for twice as many); 22 x 2 = 44. Then add this number to the number of Friday visitors; 44 + 22 = 66.
14	C	First, calculate how many hours Audrey has already watched TV by multiplying the number of shows she has watched by the length of each show; 4 x 1 = 4. Then subtract this answer from the total number of hours she is allowed to watch; 5 - 4 = 1.
15	A	First, calculate the total number of books Greg has by adding the number of books given to him by his brother to the number of books he already had; 15 + 3 = 18. Then calculate the number of books on each shelf by dividing the total number of books by the number of shelves; 18 ÷ 6 = 3.
16	A & C	There are two steps required to find the answer. First, in order to figure out the total of doughnuts multiply 2 boxes by 12 doughnuts. 12 x 2= 24. Then subtract the amount of donuts Sarah shared with friends. 24-7= 17.
17	1. C 2. B 3. D	There are three steps required to find the answer. First, in order to figure out the total amount of baseball cards, add Freddy's 32 cards to his classmate's 8 cards. 32 + 8= 40. Then to figure out the total amount of people, add Freddy to his 4 classmates. 4 + 1= 5. The last step is to divide the total amount of cards among 5 people. 40 ÷ 5= 8.
18	D	First step is total number of bottles the farmer needed: 22 ÷ 2 = 11. Next step: Number of bottles left after he spilled 6 bottles: 11 - 6 = 5. So, Option (D) is to be circled.

Question No.	Answer	Detailed Explanation			
19			$50	$5	$10	$4
	Karen had 86 dollars. He bought 7 books. After buying them he had 16 dollars. How much did each book cost?		○	○	●	○
	Jose and his four friends bought a new board game. It was on sale for 20 dollars off. If each of the boys (total 5 of them) paid $6. What was the original cost of the new board game?		●	○	○	○
	A shopkeeper buys 5 pens for $35 and sells them at the rate of $8 per pen. If he sells all the five pens, how much profit he will get?		○	●	○	○
	Jeffrey bought 8 actions figures which cost 3 dollars each from John. John bought 6 books from the amount he received from Jeffrey. If the cost of each book John purchased is the same, what is the cost of each book?		○	○	○	●

Solution to problem 1 : First, subtract $16 from $86 to get the cost of 7 books; 86 - 16 = $70. Then divide $70 by 7 to get the cost of one book. Cost of one book = 70 ÷ 7 = $10.

Solution to problem 2 : First, multiply 5 by $6 to get the total amount paid by the boys; 5 x 6 = $30. Then add $20 to $30 to get the original cost of the new board game; 20 + 30 = $50.

Solution to problem 3 : First, multiply 5 by 8 to calculate the total amount of money the shopkeeper gets; 5 x 8 = $40. Then subtract $35 from $40 to get the profit he earns; 40 - 35 = $5.

Solution to problem 4 : First, multiply 8 by $3 to get the amount received by John; 8 x 3 = $24. Then divide $24 by 6 to get the cost of each book; 24 ÷ 6 = $4.

Lesson 3: Understanding Multiplication

Question No.	Answer	Detailed Explanation
1	D	The picture depicts 3 sets of 9 objects which is equivalent to 3 x 9 = 27.
2	C	The picture depicts 4 sets of 5 objects which is equivalent to 4 x 5.
3	D	The picture depicts 4 sets of 7 objects which is equivalent to 4 x 7 = 28.
4	A	The picture depicts 2 sets of 12 objects which is equivalent to 2 x 12 = 24.
5	A	The picture depicts 4 sets of 4 objects which is equivalent to 4 x 4 = 16.
6	A	The picture depicts 4 sets of 6 objects which is equivalent to 4 x 6 = 24.
7	D	The picture depicts 3 sets of 6 objects which is equivalent to 6 x 3 (or 3 x 6) = 18
8	D	The picture depicts 5 sets of 3 objects which is equivalent to 3 x 5 (or 5 x 3) = 15.
9	A	The picture depicts 2 sets of 3 objects which is equivalent to 3 x 2 (or 2 x 3) = 6.
10	C	The picture depicts 4 sets of 5 objects which is equivalent to 5 x 4 (or 4 x 5) = 20.
11	B	The picture depicts 4 sets of 2 objects which is equivalent to 4 x 2 = 8.
12	B	The picture depicts 6 sets of 8 objects which is equivalent to 6 x 8 = 48.
13	C	The picture depicts 2 sets of 10 objects which is equivalent to 2 x 10 = 20.
14	A	The picture depicts 5 sets of 5 objects which is equivalent to 5 x 5 = 25.
15	D	The picture depicts 3 sets of 3 objects which is equivalent to 3 x 3 = 9.

Question No.	Answer	Detailed Explanation
16	4x8	The expression 8 + 8 + 8 + 8 has the same value as 4 x 8. Multiplication problems can be solved using repeated addition. Adding 4 groups of 8 is the same as multiplying 4 groups of 8.
17		

	Column A: 3+3+3+3+3+3	Column B: 3+3+3+3+3+3+3+3	Column C: 3+3+3
3 x 8	○	●	○
3 x 3	○	○	●
3 x 6	●	○	○

Question No.	Answer	Detailed Explanation
18	2x5 3x6 2x4	2 groups of 5 objects represents the expression 2 x 5. 3 groups of 6 objects represents the expression 3 x 6. 2 groups of 4 objects represents the expression 2 x 4.
19		Yes, John is correct. 8 x 6 = 8 x (5 + 1). Then John used the distributive property. 8 x (5 + 1) = 8 x 5 + 8 x 1 = 8 x 5 + 8.
20		Let n be the total number of pens the boys buy all together. n = (number of pens each boy buys) x (number of boys) = 6 x 7 = 42 pens
21		

Number of lions	5	6	9	8	4
Total number of legs	20	24	36	32	16

Lesson 4: Understanding Division

Question No.	Answer	Detailed Explanation
1	D	There are 100 items that need to be divided into 10 groups. 100 ÷ 10 = 10.
2	C	There are 15 items that need to be divided into 3 groups. 15 ÷ 3 = 5.
3	A	There are 50 items that need to be divided into 5 groups. 50 ÷ 5 = 10.
4	D	There are 30 items that need to be sorted into 6 groups. 30 ÷ 6 = 5.
5	B	There are 80 items that need to be sorted into groups of 10. 80 ÷ 10 = 8.
6	B	There are 15 items that need to be shared with 5 groups. 15 ÷ 5 = 3.
7	D	There are 81 items that need to be divided into 9 groups. 81 ÷ 9 = 9.
8	C	There are 40 items that need to be sorted into groups of 10. 40 ÷ 10 = 4.
9	B	There are 20 items that need to be sorted into groups of 5. 20 ÷ 5 = 4.
10	A	There are 30 people who need to be divided into 5 groups. 30 ÷ 5 = 6.
11	D	There are 24 items that need to be sorted into groups of 3. 24 ÷ 3 = 8.
12	B	There are 44 items that need to be divided into 4 groups. 44 ÷ 4 = 11.
13	C	All together, the group has 32 + 4 marbles which equals 36. There are 4 people in the group. There are 36 items that need to be shared with 4 groups. 36 ÷ 4 = 9.
14	B	There are 100 items (dollars) that need to be divided into 5 groups. $100 ÷ 5 = $20.
15	D	There are 26 items that need to be shared equally between two people. 26 ÷ 2 = 13.
16	4 slices	There is a total of 8 slices to be divided among 2 people. 8 ÷ 2 = 4.
17	A & C	The correct answers are A and C. There is a total of 16 stickers to be divided into equal groups. The stickers can be divided into 2 equal groups of 8. The stickers can also be divided into 4 equal groups of 4.
18	A	Picture A represents the expression 10 ÷ 5. In the picture, there is a total of 10 objects divided into 5 equal groups.

Question No.	Answer	Detailed Explanation		
19	5; 10; 50	<table><tr><th>Dividend</th><th>Possible Divisors</th></tr><tr><td>100</td><td>2</td></tr><tr><td>100</td><td>4</td></tr><tr><td>100</td><td>**5**</td></tr><tr><td>100</td><td>**10**</td></tr><tr><td>100</td><td>20</td></tr><tr><td>100</td><td>25</td></tr><tr><td>100</td><td>**50**</td></tr></table>		
20			<	>	=
		30 ÷ 5 ___ 42 ÷ 6	●	○	○
		72 ÷ 8 ___ 63 ÷ 7	○	○	●
		54 ÷ 6 ___ 56 ÷ 7	○	●	○

Lesson 5: Multiplication & Division Facts

Question No.	Answer	Detailed Explanation
1	C	In multiplication, if one of the factors is 0, the product is also 0.
2	C	The Identity Property of Multiplication states that any number multiplied by 1 equals itself, number so 1 x 10 = 10.
3	A	3 x 8 represents 3 groups of 8 items. There are 24 items in total.
4	B	5 x 9 represents 5 groups of 9 items. There are 45 items in total
5	C	Product refers to the answer when numbers are multiplied. 8 x 6 represents 8 groups of 6 items. There are 48 items in total.
6	C	Product refers to the answer when numbers are multiplied. 7 x 7 represents 7 groups of 7 items. There are 49 items in total.
7	B	Product refers to the answer when numbers are multiplied. 4 x 6 represents 4 groups of 6 items. There are 24 items in total.
8	A	Product refers to the answer when numbers are multiplied. 6 x 9 represents 6 groups of 9 items. There are 54 items in total.
9	B	Product refers to the answer when numbers are multiplied. 9 x 8 represents 9 groups of 8 items. There are 72 items in total.
10	D	Product refers to the answer when numbers are multiplied. Option D is the only choice in which the answer is 48.
11	B	The quotient refers to the answer when a number is divided by another number. There are 25 items that need to be divided into 5 groups. 25 ÷ 5 = 5.
12	B	There are 32 items that need to be divided into 4 groups. 32 ÷ 4 = 8
13	A	There are 28 items that need to be divided into 7 groups. 28 ÷ 7 = 4. The quotient refers to the answer when a number is divided by another number.
14	A	When the number 0 is divided by any non-zero number, the answer is always 0. The quotient refers to the answer when a number is divided by another number.
15	C	When a number is divided by 1, the answer is always the original number.
16	D	The quotient refers to the answer when a number is divided by another number. There are 12 items that need to be divided into 2 groups. 12 ÷ 2 = 6.
17	B	There are 63 items that need to be divided into 9 groups. 63 ÷ 9 = 7.

LumosLearning.com

Question No.	Answer	Detailed Explanation
18	B	There are 42 items that need to be divided into 7 groups. 42 ÷ 7 = 6
19	A	The quotient refers to the answer when a number is divided by another number. There are 33 items that need to be divided into 3 groups. 33 ÷ 3 = 11.
20	C	There are 56 items that need to be divided into 7 groups. 56 ÷ 7 = 8.
21	B	4 x 12 represents 4 groups of 12 items. There are 48 items total.
22	B	There are 75 items that need to be divided into 5 groups. 75 ÷ 5 = 15
23	A	There are 84 items that need to be divided into 12 groups. 84 ÷ 12 = 7.
24	B	12 x 3 represents 12 groups of 3 items. There are 36 items total.
25	B	There are 36 items that need to be divided into 3 groups. 36 ÷ 3 = 12.
26	C	There are 60 items that need to be divided into 5 groups. 60 ÷ 5 = 12.
27	B	11 x 4 represents 11 groups of 4 items. There are 44 items total.
28	A	There are 80 items that need to be divided into 8 groups. 80 ÷ 8 = 10.
29	D	12 x 8 represents 12 groups of 8 items. There are 96 items total.
30	C	There are 50 items that need to be divided into 5 groups. 50 ÷ 5 = 10.

31

	12	18	32
6 x 3 =	○	●	○
8 x 4 =	○	○	●
4 x 3 =	●	○	○
9 x 2 =	○	●	○

Question No.	Answer	Detailed Explanation
32	6	The quotient is the result of dividing a number by another number. The quotient of 24 ÷ 4 is 6.
33	A & D	Product refers to the answer when numbers are multiplied. 6 x 6 = 36 and 9 x 4 = 36. Therefore, options (A) and (D) are correct.
34	C	The quotient refers to the answer when a number is divided by another number. There are 27 items that need to be divided into 9 groups. 27 ÷ 9 = 3.

35

5	x	8	=	40
8	÷	1	=	8
0	÷	7	=	0
6	x	5	=	30

Lesson 6: Multiplication & Division Properties

Question No.	Answer	Detailed Explanation
1	D	In order for the statement to be true, the answer on both sides of the equal sign must be the same. All of the answer choices are equal except for the last choice. 12 x 1 = 12 and 12 x 12 = 144. 12 x 1 is not equal to 12 x 12.
2	A	Option A is the only one that is true because 11 x 6 = 66 and 6 x 11 = 66. This is an example of the Commutative Property of Multiplication.
3	C	In multiplication, the only time that the number 0 will be the product is when at least one of the factors is 0. Option C is the only choice that fits this rule.
4	C	In multiplication, the only time that the number 0 will be the product is when at least one of the factors is 0. In division, the only time that the number 0 will be the answer is when the dividend is 0. Option C is the only choice where the equation fits this rule.
5	C	The Identity Property of Multiplication states that any number multiplied by 1 equals itself.
6	A	The Commutative Property of Multiplication states that the order of the factors does not change the answer.
7	B	The Associative Property of Multiplication states that the grouping of factors does not change the answer.
8	D	The Distributive Property states that multiplying a number by a group of numbers added together is the same as doing each multiplication problem separately.
9	C	The Commutative Property of Multiplication states that the order of the factors does not change the answer. 4 x 5 = 20 is the same as 5 x 4 = 20.
10	A	The Associative Property of Multiplication states that the grouping of factors does not change the answer. So (2 x 3) x 4 = 24 is the same as 2 x (3 x 4) = 24.
11	C	Inverse operations mean that one operation will reverse the effect of another. Division is the inverse of multiplication and vice versa. For example, if 4 x 3 = 12 then 12 ÷ 3 = 4 and 12 ÷ 4 = 3.
12	A	The Commutative Property of Multiplication states that the order of the factors does not change the answer.
13	D	The Distributive Property states that multiplying a number by a group of numbers added together is the same as doing each multiplication separately.
14	C	The Identity Property of Multiplication states that any number multiplied by 1 equals itself.

Question No.	Answer	Detailed Explanation
15	A	The Identity Property of Multiplication states that any number multiplied by 1 equals itself. 10 x 1 = 10.
16	C & D	Commutative property is the rule that states a x b = b x a. According to this rule, 4 x 5 = 5 x 4.
17	1	Identity property is the rule that states that when a number is multiplied by 1, the product, is the original number. This applies to the equation 7 x 1 = 7.
18	A	Associative property is the rule that states that when the grouping of factors changes, the product remains the same. This applies to the equations (2 x 3) x 4 = 24 and 2 x (3 x 4) = 24.
19 Part A	3	The commutative property of multiplication states that the order of the factors does not change the answer. Here, 3 and 7 are the factors on the left side. Therefore, the unknown factor on the right side has to be 3.
19 Part B	Commutative property	The answer is commutative property. The property states that a x b = b x a.

Question 20:

	3 x (5 x 7) = (3 x 5) x 7	3 x 1 = 3	3 x 5 = 5 x 3	3 x (5 + 7) = (3 x 5) + (3 x 7)
Commutative Property	○	○	●	○
Associative Property	●	○	○	○
Identity Property	○	●	○	○
Distributive Property	○	○	○	●

Lesson 7: Applying Multiplication & Division

Question No.	Answer	Detailed Explanation
1	B	Product refers to the result of the multiplication of two or more numbers. 54 and 3 are both factors.
2	D	The phrase "twice as many" indicates that if a number is multiplied by 2, the product will reflect two times, or twice, the original amount. In this case, the product of 30 popcorn balls is already known. The product must be then divided by 2 in order to find the number of cotton candies. 30÷2=15.
3	C	The phrase "8 times as many" indicates that if Sue's amount of DVDs is multiplied by 8, the product will be equal to the amount of Monica's DVDs. To solve for Sue use the equation n x 8 = 56. When trying to solve for a missing number in a multiplication equation, you must divide the product by the given number. 56÷8=7
4	A	Marcus' jumping jack is equivalent to 4 times that of Jonathan. 7 x 4= 28.
5	D	Marsha's movie count is equivalent to 3 times that of Jonathan's. 3 x 4= 12.
6	B	There are 4 groups and each group has 9 items. This indicates that if the number of groups is multiplied by the number of items in each group, the product will reflect the total number of items in all. 4 x 9= 36.
7	B	The number of students can be divided by the number of slices in each pizza. The quotient will reflect the number of pizzas needed. 24 ÷ 12 = 2 pizzas.
8	C	The total number of apples can be divided by the number of apples needed for each pie. The quotient will reflect the number of pies that can be made. 27 ÷ 3= 9.
9	C	The number of rows multiplied by the number of seeds in each row must equal 48. All answer choices given contain two numbers with a product of 48 except Option C. 7 x 7 = 49.
10	B	The total number of students can be divided by the number of teams. The quotient will reflect the number of students on each team. 25 ÷ 5= 5.
11	A	The number of chapters can be divided by the number of days. The quotient will reflect the number of chapters that can be read each day. 21 ÷ 7 = 3.

Question No.	Answer	Detailed Explanation
12	D	The $40 Devon has can be divided by the cost of one gallon of fuel ($5). The quotient will reflect the number of gallons Devon can afford to buy. $40 ÷ $5 = 8 gallons.
13	B	There are 4 cups needed for one batch. Amanda is making 3 batches, so she needs 4 sets of 3 cups or 4 x 3 = 12 cups.
14	C	The phrase "twice as many" indicates that if a number is multiplied by 2 the product will reflect two times, or twice, the original amount. The amount of friends invited (20) must be multiplied by 2 in order to find out how many friends attended. The correct number sentence is 20 x 2 = 40.
15	C	Product refers to multiplication. The problem states that 9 and a number (n) when multiplied together equals 45. The correct number sentence is 9 x n = 45.
16	B & C	The correct number sentences are 5 x 6= 30 and 6 x 5= 30. There are 5 groups of 6 desks. In order to find the total number of desks, multiply 5 by 6. The rule of communicative property states that 5 x 6=6 x 5. Therefore 6 x 5= 30 is correct as well.
17	C	Each friend should receive 6 flowers. There are 24 flowers to be divided among 4 people. In order to find the number of flowers per person, divide 24 by 4. 24 ÷ 4= 6.
18	8	This is a problem on division. Number of cupcakes each boy gets = Total number of cupcakes ÷ number of boys= 48 ÷ 6 = 8 cupcakes
19	A) x B) +	A) First, we find the value of 64 ÷ 8. 64 ÷ 8 = 8. Now, we have to find the correct symbol to make the equation 8 = 2 ___ 4, true. Multiplying 2 by 4, we get 8. Therefore, x (multiplication symbol) is the correct choice. B) First, we find the value of 42 ÷ 7. 42 ÷ 7 = 6. Now, we have to find the correct symbol to make the equation 2 ___ 4 = 6, true. If we add 2 and 4, we get 6. Therefore, + (plus) is the correct choice.
20		Let n be the number of days Joseph takes to complete the book, p be the number of pages in the book (p = 56) and r be the number of pages he reads every day (r = 8). n = p ÷ r = 56 ÷ 8 = 7 days

Lesson 8: Number Patterns

Question No.	Answer	Detailed Explanation
1	C	An even number is any number whose ones digit is one of the following numbers: 0, 2, 4, 6, 8. Option C is the only choice that fits this criteria.
2	C	An odd number is any number whose ones digit is one of the following numbers: 1, 3, 5, 7, 9. Option C is the only choice that does not contain any numbers that fit this criteria.
3	C	The rule states that when two even numbers are added, the answer will always be even. For example, 34 + 12 = 46.
4	A	The rule states that when two even numbers are multiplied, the product will always be even. For example, 34 x 4 = 136.
5	D	There is not enough information given for us to decide if the number is a multiple of 3, 7, or 9. For example, if the original number was 19, it would not be a multiple of 3, 7, or 9.
6	D	Any multiple of an even number is also even. Numbers that are multiples of 8 are also multiples of 2 and 4 because 2 and 4 are factors of 8.
7	D	The pattern is adding 7 to each number to make the next number (7 + 7 = 14, 14 + 7 = 21 and so on). The next number will be 35 + 7 = 42, the next will be 42 + 7 = 49, and the next will be 49 + 7 = 56. These numbers also represent the multiples of 7 in order.
8	A	"Two equal addends" means the number can be divided into two equal numbers. This can be performed on all even numbers. The number 6 is an even number and all of its multiples are also even.
9	B	The pattern is adding 9 to each number to make the next number (9 + 9 = 18, 18 + 9 = 27 and so on). The next number will be 36 + 9 = 45, the next will be 45 + 9 = 54, and the next will be 54 + 9 = 63.
10	C	If an odd number is multiplied by an even number, the answer will be an even number.
11	B	Multiples of even numbers are always even. Four is an even number, so all of its multiples are also even.
12	B	The rule states that when an odd number is added to an even number, the answer will always be odd. For example, 45 + 18 = 63.
13	D	The multiples of 4 are {4, 8, 12, 16, 20, 24, 28, 32, . . .} It is apparent from this list that a multiple of 4 could have a 6, 8, or 2 as its ones digit.
14	A	All multiples of 5 have a number 0 or 5 in the ones position. For example, 205 and 210 are both multiples of 5.
15	D	Whenever an even number is multiplied by any number, the answer will always be even.

Question No.	Answer	Detailed Explanation
16	<table><tr><td></td><td>10</td><td>35</td><td>24</td></tr><tr><td>2, 4, 6, 8</td><td>●</td><td>○</td><td>○</td></tr><tr><td>40, 36, 32, 28</td><td>○</td><td>○</td><td>●</td></tr><tr><td>7, 14, 21, 28</td><td>○</td><td>●</td><td>○</td></tr></table>	If the pattern in the first row continues to increase by 2, the next number will be 10. If the pattern in the second row continues to decrease by 4, the next number will be 24. If the pattern in the last row continues to increase by 7, the next number will be 35.
17	12; 6; 21; 8; 24	<table><tr><th>IN</th><th>OUT</th></tr><tr><td>3</td><td>9</td></tr><tr><td>4</td><td>**12**</td></tr><tr><td>5</td><td>15</td></tr><tr><td>**6**</td><td>18</td></tr><tr><td>7</td><td>**21**</td></tr><tr><td>**8**</td><td>**24**</td></tr></table> If the pattern continues to increase by multiples of 3, the missing numbers will be 12, 6, 21, 8 and 24.
18	A & D	If the pattern continues to decrease by 10, the only numbers that will appear from the answer choices are 50 and 60.
19	C	When you add two odd numbers, the sum is an even number. Addition and subtraction are inverse operations. Therefore, when you subtract an odd number from an even number, the difference has to be an odd number. If a and b are odd numbers, the sum (c = a + b) is an even number. Therefore, c - a (= b) or c - b (= a) has to be an odd number.
20		

	2	4	5	7
A number has a four in its ones place. The number can be a multiple of _____.	●	●	○	●
A number has a five in its ones place. The number can be a multiple of _____.	○	○	●	●
A number has a zero in its ones place. The number can be a multiple of _____.	●	●	●	●
A number has a three in its ones place. The number can be a multiple of _____.	○	○	○	●

Statement 1: A multiple of 5 cannot have four in the ones place. A multiple of 2 or 4 or 7 can have four in the ones place. For ex. 2 x 2 = 4, 4 x 6 = 24, 2 x 7 = 14.
Statement 2: A multiple of 2 or 4 cannot have five in the ones place. A multiple of 5 or 7 can have five in the ones place. For ex. 5 x 7 = 35.
Statement 3: Any number when multiplied by 10 results in a number with a zero in the ones place. For ex. 2 x 10 = 20, 4 x 10 = 40, 5 x 10 = 50, 7 x 10 = 70.
Statement 4: Multiples of 2, or 4, or 5 cannot have three in the ones place. A multiple of 7 can have 3 in the ones place. For ex. 7 x 9 = 63.

Date of Completion:_____   Score:_____

Chapter 3: Geometry

Lesson 1: 2-Dimensional Shapes

1. Fill in the blank with the correct term.
   Closed, plane figures that have straight sides are called _____ .

   Ⓐ parallelograms
   Ⓑ line segments
   Ⓒ polygons
   Ⓓ squares

2. Which of the following shapes is not a polygon?

   Ⓐ Square
   Ⓑ Hexagon
   Ⓒ Circle
   Ⓓ Pentagon

3. Complete this statement.
   A rectangle must have _____ .

   Ⓐ four right angles
   Ⓑ four straight angles
   Ⓒ four obtuse angles
   Ⓓ four acute angles

4. How many sides does a trapezoid have?

   Ⓐ 4
   Ⓑ 8
   Ⓒ 6
   Ⓓ 10

5. Complete the following statement.
   A square is always a _____ .

   Ⓐ rhombus
   Ⓑ parallelogram
   Ⓒ rectangle
   Ⓓ All of the above

6. Which of these statements is true?

    Ⓐ A square and a triangle have the same number of angles.
    Ⓑ A triangle has more angles than a square.
    Ⓒ A square has more angles than a triangle.
    Ⓓ A square and a triangle each have no angles.

7. Which of these statements is true?

    Ⓐ A rectangle has more sides than a trapezoid.
    Ⓑ A parallelogram and a trapezoid have the same number of sides.
    Ⓒ A triangle has more sides than a trapezoid.
    Ⓓ A triangle has more sides than a square.

8. Complete this statement.
    A trapezoid must have _____.

    Ⓐ two acute angles
    Ⓑ two right angles
    Ⓒ one pair of parallel sides
    Ⓓ two pairs of parallel sides

9. Complete the following statement.
    Squares, rectangles, rhombi and trapezoids are all _____.

    Ⓐ triangles
    Ⓑ quadrilaterals
    Ⓒ angles
    Ⓓ round

10. Which of these shapes is a quadrilateral?

    Ⓐ circle
    Ⓑ triangle
    Ⓒ rectangle
    Ⓓ pentagon

11. Which of these shapes is NOT a quadrilateral?

    Ⓐ square
    Ⓑ trapezoid
    Ⓒ rectangle
    Ⓓ triangle

12. Name the figure shown below.

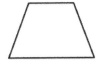

Ⓐ Trapezoid
Ⓑ Square
Ⓒ Pentagon
Ⓓ Rhombus

13. Name the object shown below.

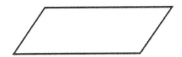

Ⓐ Rectangle
Ⓑ Parallelogram
Ⓒ Trapezoid
Ⓓ Rhombus

14. The figure shown below is a _____ .

Ⓐ parallelogram
Ⓑ rectangle
Ⓒ quadrilateral
Ⓓ All of the above

15. The figure below is a _____ .

Ⓐ triangle
Ⓑ square
Ⓒ rhombus
Ⓓ trapezoid

16. Are these figures quadrilaterals? Select yes or no.

	Yes	No
Circle		
Star		
Square		
Rectangle		

17. Circle the parallelogram.

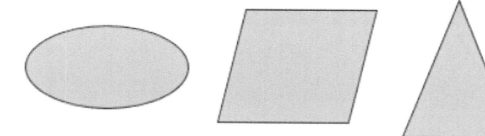

18. For each polygon in the first column, an attribute is defined in the second column. Write true, if the polygon has the mentioned attribute or write false if the polygon does not have the mentioned attribute.

Polygon	Attribute	True or False
Rhombus	It has two sets of parallel sides	True
Parallelogram	All the angles are equal	
Rectangle	Opposite sides are equal	

19. Draw a quadrilateral which has three obtuse angles.
    Instruction : An obtuse angle is an angle which measures more than 90° but less than 180°.

20. Which of the following figures have at least one set parallel sides? Note that more than one option may be correct.

Ⓐ

Ⓑ

Ⓒ

Ⓓ

CHAPTER 3 → Lesson 2: Shape Partitions

1. What is the dotted line that divides a shape into two equal parts called?

   Ⓐ a middle line
   Ⓑ a line of symmetry
   Ⓒ a line of congruency
   Ⓓ a divider

2. A square has how many lines of symmetry?

   Ⓐ 8
   Ⓑ 4
   Ⓒ 1
   Ⓓ 2

3. Which of the following has NO lines of symmetry?

   Ⓐ

   Ⓑ

   Ⓒ

   Ⓓ

4. Which of the following objects has more than one line of symmetry?

Ⓐ

Ⓑ

Ⓒ

Ⓓ

5. What fraction of this triangle is shaded?

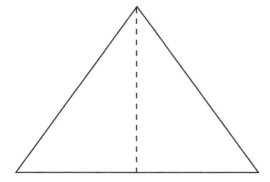

Ⓐ $\frac{1}{2}$

Ⓑ $\frac{2}{2}$

Ⓒ $\frac{0}{2}$

Ⓓ $\frac{3}{4}$

6. What fraction of this triangle is shaded?

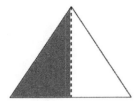

Ⓐ $\dfrac{0}{2}$

Ⓑ $\dfrac{1}{2}$

Ⓒ $\dfrac{2}{2}$

Ⓓ $\dfrac{3}{4}$

7. What fraction of this triangle is shaded?

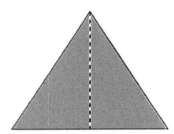

Ⓐ $\dfrac{0}{2}$

Ⓑ $\dfrac{1}{2}$

Ⓒ $\dfrac{2}{2}$

Ⓓ $\dfrac{3}{4}$

8. What fraction of this square is shaded?

Ⓐ $\dfrac{0}{2}$

Ⓑ $\dfrac{1}{2}$

Ⓒ $\dfrac{2}{2}$

Ⓓ $\dfrac{3}{4}$

9. What fraction of this square is shaded?

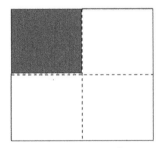

Ⓐ $\dfrac{0}{4}$

Ⓑ $\dfrac{1}{4}$

Ⓒ $\dfrac{2}{4}$

Ⓓ $\dfrac{1}{2}$

10. What fraction of this square is shaded?

Ⓐ $\frac{0}{4}$

Ⓑ $\frac{1}{4}$

Ⓒ $\frac{1}{2}$

Ⓓ $\frac{3}{4}$

11. What fraction of this rectangle is shaded?

Ⓐ $\frac{0}{4}$

Ⓑ $\frac{1}{4}$

Ⓒ $\frac{2}{4}$

Ⓓ $\frac{3}{4}$

12. What fraction of this circle is shaded?

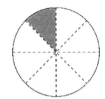

Ⓐ $\dfrac{1}{8}$

Ⓑ $\dfrac{1}{4}$

Ⓒ $\dfrac{1}{2}$

Ⓓ $\dfrac{3}{4}$

13. What fraction of this circle is shaded?

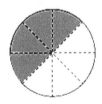

Ⓐ $\dfrac{1}{8}$

Ⓑ $\dfrac{1}{4}$

Ⓒ $\dfrac{5}{8}$

Ⓓ $\dfrac{4}{8}$

14. The area of the entire rectangle shown below is 48 square feet. What is the area of the shaded portion?

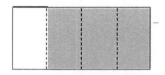

Ⓐ 36 square feet
Ⓑ 48 square feet
Ⓒ 144 square feet
Ⓓ 12 square feet

15. If the area of the entire rectangle below is 36 square feet. What is the area of the shaded portion?

Ⓐ 8 square feet
Ⓑ 9 square feet
Ⓒ 144 square feet
Ⓓ 12 square feet

16. Do these figures have a line of symmetry? Select yes or no.

	Yes	No
♥		
▼		
◔		

17. Circle the shape that has a line of symmetry.

18. If the area of the whole rectangle is 28 square units, what is the area of the shaded portion? Write your answer in the box given below.

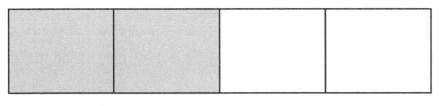

square units

19. Shade one third of the figure below.

20. A circle has an area of 96 sq. cm. The circle is divided into 8 equal parts. Which of the following statements are correct? Select all the correct answers.

Ⓐ If you shade 3 parts, the area of the shaded portion is 32 sq. cm.
Ⓑ If you shade 4 parts, the area of the shaded portion is 48 sq. cm.
Ⓒ If you shade 7 parts, the area of the shaded portion is 84 sq. cm.
Ⓓ If you shade 2 parts, the area of the shaded portion is 24 sq. cm.

End of Geometry

ANSWER KEY AND DETAILED EXPLANATION

Chapter 3: Geometry
Lesson 1: 2-Dimensional Shapes

Question No.	Answer	Detailed Explanation
1	C	By definition, a polygon is a plane (flat), closed figure with only straight sides.
2	C	A polygon must have only straight sides. A circle is the only option that does not fit this criteria. Since it is curved.
3	A	A rectangle is a quadrilateral (4-sided polygon) with 4 right angles.
4	A	A trapezoid is a quadrilateral which means it has 4 sides.
5	D	A square is a rhombus parallelogram, and rectangle because all of these figures are four-sided and contain two sets of parallel sides.
6	C.	Option C is true because a square has 4 angles and a triangle has 3 angles.
7	B	Option B is true because a parallelogram and a trapezoid both have 4 sides.
8	C	A trapezoid is a quadrilateral with one pair of parallel sides.
9	B	A quadrilateral is a figure with four straight sides and four angles. Squares, rectangles, rhombi, and trapezoids all have 4 sides.
10	C	A quadrilateral is a figure with four straight sides and four angles. A rectangle fits this description, whereas a triangle has 3 sides, a circle is round, and a pentagon has 5 sides.
11	D	A quadrilateral is a figure with four straight sides and four angles. A triangle is the only choice that does not fit this description. Since it has only 3 sides.
12	A	A trapezoid is a quadrilateral that contains only one pair of parallel sides.
13	B	A parallelogram is a quadrilateral with two pairs of opposite parallel sides. The figure is not a rectangle because it does not have right angles. It is not a rhombus because the sides are not all equal in length. It is not a trapezoid because a trapezoid only has one set of parallel sides.
14	D	The shape is a parallelogram because it has two pairs of parallel sides. It is a quadrilateral because it has 4 sides. It is a rectangle because it is a parallelogram with all right angles.
15	C	A rhombus is a quadrilateral with 4 equal sides. This figure is not a square because it does not have right angles. Triangles are three-sided, while trapezoids do not have four equal sides.

Question No.	Answer	Detailed Explanation
16		

	Yes	No
Circle		✓
Star		✓
Square	✓	
Rectangle	✓	

A quadrilateral is a four-sided polygon with four angles. Squares and rectangles belong in this category. Circles and stars are not quadrilaterals.

17		A parallelogram is a (non-self-intersecting) quadrilateral with two pairs of parallel sides. Ovals and triangles are not parallelograms.

18		

Polygon	Attribute	True or False
Rhombus	It has two sets of parallel sides	True
Parallelogram	All the angles are equal	**False**
Rectangle	Opposite sides are equal	**True**

In a parallelogram, opposite angles are equal. Adjacent angles need not be equal.

A rectangle is a special type of parallelogram, whose angles measure 90° each. Since a rectangle is a type of parallelogram, it has the attribute: opposite sides are equal.

19		In the above quadrilateral, angles ADC, DAB and ABC are obtuse angles.
20	C & D	The first figure is a trapezoid. It has one pair of parallel sides. The second figure is a pentagon. It has no parallel sides. The third figure is a regular hexagon. It has 3 sets of parallel sides. The fourth figure has one set of parallel sides.

Lesson 2: Shape Partitions

Question No.	Answer	Detailed Explanation
1	B	A line of symmetry is an imaginary line that divides an object into two mirror images.
2	B	A line of symmetry is an imaginary line that divides an object into two mirror images. A square can be divided across the length, across the width, down diagonally from left to right, and down diagonally from right to left.
3	C	A line of symmetry is an imaginary line that divides an object into two mirror images. Option C cannot be divided in such a way.
4	A	A line of symmetry is an imaginary line that divides an object into two mirror images. The object in Option A can have a line of symmetry at multiple points, for example across the length, across the width, and diagonally.
5	C	No parts of the triangle are shaded yet the shape is divided into two parts. To form a fraction, the numerator is the part and the denominator is the whole. Since no parts are shaded, the numerator would be 0.
6	B	One part of the triangle is shaded yet the shape is divided into two parts. To form a fraction, the numerator is the part and the denominator is the whole.
7	C	Two parts of the triangle are shaded and the shape is divided into two parts. To form a fraction, the numerator is the part and the denominator is the whole.
8	B	One part of the square is shaded and the shape is divided into two parts. To form a fraction, the numerator is the part and the denominator is the whole.
9	B	One part of the square is shaded and the shape is divided into four parts. To form a fraction, the numerator is the part and the denominator is the whole.
10	D	Three parts of the square are shaded and the shape is divided into four parts. To form a fraction, the numerator is the part and the denominator is the whole.
11	B	One part of the rectangle is shaded yet the shape is divided into four parts. To form a fraction, the numerator is the parts and the denominator is the whole.
12	A	One part of the circle is shaded yet the shape is divided into eight parts. To form a fraction, the numerator is the parts and the denominator is the whole.

Question No.	Answer	Detailed Explanation
13	D	Four parts of the circle are shaded yet the shape is divided into eight parts. To form a fraction, the numerator is the parts and the denominator is the whole.
14	A	The shape is divided into four equal parts. This means that each part has the same area. If the total area is known, divide this area by 4 to calculate the area of each part. 48 ÷ 4 = 12. Since one part is left unshaded, subtract 12 from the total area of 48 to find that the shaded portion represents 36 square feet.
15	B	The shape is divided into four equal parts. This means that each part has the same area. If the total area is known, divide this area by 4 to calculate the area of each part. 36 ÷ 4 = 9. Each piece has an area of 9 square feet. The problem asks for the area of the shaded portion. There is only one shaded section, so the area equals 9 square feet.
16		

	Yes	No
♥	✓	
▽	✓	
◔	✓	

A figure that can reflect over a line and appear unchanged has reflection symmetry or line symmetry. All 3 figures have a line of symmetry.

Question No.	Answer	Detailed Explanation
17	Star	A figure that can reflect over a line and appear unchanged has reflection symmetry or line symmetry. The star has a line of symmetry.
18	14	The shapes are divided into 4 equal parts. Each part has the same area. Since the area of the whole figure is 28, divide 28 by 4. The area of each shaded part is 7. There are 2 shaded parts so add 7 + 7. The area of the shaded portion is 14.
19	4 cells	The figure is divided into 12 equal parts. One third of 12 means we have to divide 12 by 3; 12 ÷ 3 = 4. So, we have to shade 4 cells.

Question No.	Answer	Detailed Explanation
20	B, C & D	The circle is divided into 8 equal parts. This means that each part has the same area. Divide the total area by 8 to calculate the area of each part; 96 ÷ 8 = 12 sq. cm. Each part has an area of 12 sq. cm. (1) When we shade 3 parts, the shaded portion has an area of 3 x 12 = 36 sq. cm. Therefore, option (A) is wrong. (2) When we shade 4 parts, the shaded portion has an area of 4 x 12 = 48 sq. cm. Therefore, option (B) is correct. (3) When we shade 7 parts, the shaded portion has an area of 7 x 12 = 84 sq. cm. Therefore, option (C) is correct. (4) When we shade 2 parts, the shaded portion has an area of 2 x 12 = 24 sq. cm. Therefore, option (D) is correct.

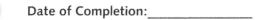

Date of Completion:_____ Score:_____

Chapter 4: Measurement

Lesson 1: Liquid Volume & Mass

1. "40 pounds" is printed at the bottom of a bag of sand. The number "40" is being used to _____ .

   Ⓐ count
   Ⓑ name
   Ⓒ locate
   Ⓓ measure

2. In the metric system, which is the best unit to measure the mass of a coffee table?

   Ⓐ Milliliters
   Ⓑ Kilograms
   Ⓒ Grams
   Ⓓ Liters

3. Which unit should be used to measure the amount of water in a small bowl?

   Ⓐ Cups
   Ⓑ Gallons
   Ⓒ Inches
   Ⓓ Tons

4. Which unit in the customary system is best suited to measure the weight of a coffee table?

   Ⓐ Gallons
   Ⓑ Pounds
   Ⓒ Quarts
   Ⓓ Ounces

5. Which of these units could be used to measure the capacity of a container?

   Ⓐ pints
   Ⓑ feet
   Ⓒ pounds
   Ⓓ millimeters

6. Which of these is a unit of mass?

   Ⓐ liter
   Ⓑ meter
   Ⓒ gram
   Ⓓ degree

7. Which of these units has the greatest capacity?

    Ⓐ gallon
    Ⓑ pint
    Ⓒ cup
    Ⓓ quart

8. Which of these might be the weight of an average sized 8 year-old child?

    Ⓐ 15 pounds
    Ⓑ 30 pounds
    Ⓒ 65 pounds
    Ⓓ 150 pounds

9. Volume is measured in _____ units.

    Ⓐ cubic
    Ⓑ liters
    Ⓒ square
    Ⓓ box

10. What is an appropriate unit to measure the weight of a dog?

    Ⓐ tons
    Ⓑ pounds
    Ⓒ inches
    Ⓓ gallons

11. What is an appropriate unit to measure the amount of water in a swimming pool?

    Ⓐ teaspoons
    Ⓑ cups
    Ⓒ gallons
    Ⓓ inches

12. What is an appropriate unit to measure the distance across a city?

    Ⓐ centimeters
    Ⓑ feet
    Ⓒ inches
    Ⓓ miles

13. What is an appropriate unit to measure the amount of salt in a cupcake recipe?

    Ⓐ teaspoons
    Ⓑ gallons
    Ⓒ miles
    Ⓓ kilograms

14. Which unit is the largest?

    Ⓐ mile
    Ⓑ centimeter
    Ⓒ foot
    Ⓓ inch

15. Which unit is the smallest?

    Ⓐ kilometer
    Ⓑ centimeter
    Ⓒ millimeter
    Ⓓ inch

16. Which units of measurement could be used to measure how much water a pot can hold? Select all the correct answers.

    Ⓐ quarts
    Ⓑ centimeter
    Ⓒ liters
    Ⓓ miles

17. Observe the figure given. How many cups of liquid does this measuring cup hold? Write your answer in the box given below.

18. Circle the tool that should be used to measure a small amount of sugar.

19. There are 8 water coolers in a school. Each water cooler holds 7 liters of water. All the water coolers were filled up in the morning. In the evening 5 liters of water remained. How much water was consumed? Explain how you got the answer in the box below.

CHAPTER 4 → Lesson 2: Measuring Length

1. Which of these units is part of the metric system?

   Ⓐ Foot
   Ⓑ Mile
   Ⓒ Kilometer
   Ⓓ Yard

2. Which metric unit is closest in length to one yard?

   Ⓐ decimeter
   Ⓑ meter
   Ⓒ millimeter
   Ⓓ kilometer

3. Which of these is the best estimate for the length of a table?

   Ⓐ 2 decimeters
   Ⓑ 2 centimeters
   Ⓒ 2 meters
   Ⓓ 2 kilometers

4. What unit should you use to measure the length of a book?

   Ⓐ Kilometers
   Ⓑ Meters
   Ⓒ Centimeters
   Ⓓ Grams

5. About how long is a new pencil?

   Ⓐ 8 inches
   Ⓑ 8 feet
   Ⓒ 8 yards
   Ⓓ 8 miles

6. Which of these is the best estimate for the length of a football?

   Ⓐ 1 foot
   Ⓑ 2 feet
   Ⓒ 6 feet
   Ⓓ 4 feet

7. Complete the following statement.
   The length of a football field is _____.

   Ⓐ less than one meter
   Ⓑ greater than one meter
   Ⓒ about one meter
   Ⓓ impossible to measure

8. Complete the following statement.
   An adult's pointer finger is about one _____ wide.

   Ⓐ meter
   Ⓑ kilometer
   Ⓒ millimeter
   Ⓓ centimeter

9. Complete the following statement.
   The distance between two cities would most likely be measured in _____.

   Ⓐ hours
   Ⓑ miles
   Ⓒ yards
   Ⓓ square inches

10. A ribbon is 25 centimeters long. About how many inches long is it?

    Ⓐ 2
    Ⓑ 25
    Ⓒ 10
    Ⓓ 50

11. _____

    How long is this object?

    Ⓐ 4 inches
    Ⓑ 8 inches
    Ⓒ 10 inches
    Ⓓ 12 inches

12. ──

How long is this object?

Ⓐ 2 inches
Ⓑ 5 inches
Ⓒ 3 inches
Ⓓ 1 inch

13.

How long is this object?

Ⓐ 6 inches
Ⓑ 5 and a half inches
Ⓒ 6 and a half inches
Ⓓ 7 inches

14. Which statement is correct?

Ⓐ 1 inch > 1 mile
Ⓑ 1 inch > 1 centimeter
Ⓒ 1 foot < 1 inch
Ⓓ 1 mile < 1 foot

15. Which is correct?

Ⓐ 12 inches > 1 foot
Ⓑ 12 inches < 1 foot
Ⓒ 12 inches = 1 foot
Ⓓ 9 inches = 1 foot

16. Which of the following could be measured with a ruler? Select all correct answers.

Ⓐ water in a bowl
Ⓑ a football field
Ⓒ a carrot
Ⓓ a crayon

17. Observe the figure. How long is the pencil when measured in inches?

18. Fill in the correct answer in the blanks shown in the table.

Measurement in inches	Measurement in half inches	Measurement in quarter inches
$3\frac{1}{2}$ inches	7 half inches	14 quarter inches
$2\frac{1}{2}$ inches		
	11 half inches	
		26 quarter inches

19. Use the line plot to answer the questions given in the first column.

Lengths of Fish

Instruction:
X = 2 fishes

	4	6	8
How many fish are $16\frac{1}{2}$ inches long?	○	○	○
How many more fish are 16 inches long than 17 inches?	○	○	○
How many fish are less than $15\frac{3}{4}$ inches long?	○	○	○

CHAPTER 4 → Lesson 3: Telling Time

1. What time does this clock show?

   Ⓐ 3:12
   Ⓑ 2:17
   Ⓒ 2:22
   Ⓓ 2:03

2. What time does this clock show?

   Ⓐ 5:42
   Ⓑ 9:28
   Ⓒ 6:47
   Ⓓ 5:47

3. What time does this clock show?

   Ⓐ 10:00
   Ⓑ 12:50
   Ⓒ 10:02
   Ⓓ 9:41

4. What time does this clock show?

Ⓐ 12:39
Ⓑ 8:04
Ⓒ 1:38
Ⓓ 12:42

5. On an analog clock, the shorter hand shows the _____ .

Ⓐ minutes
Ⓑ hours
Ⓒ seconds
Ⓓ days

6. On an analog clock, the longer hand shows the _____ .

Ⓐ minutes
Ⓑ hours
Ⓒ days
Ⓓ seconds

7. The clock currently shows:

   What time will it be in 8 minutes?

   Ⓐ 1:38
   Ⓑ 10:15
   Ⓒ 10:10
   Ⓓ 12:58

8. The clock currently shows:

   What time will it be in 20 minutes?

   Ⓐ 12:59
   Ⓑ 1:09
   Ⓒ 2:00
   Ⓓ 8:24

9. The clock says:

   What time was it 10 minutes ago?

   Ⓐ 1:29
   Ⓑ 12:29
   Ⓒ 12:49
   Ⓓ 1:09

10. Lucy started her test at 12:09 PM and finished at 12:58 PM. David started at 12:15 PM and ended at 1:03 PM. Who finished in a shorter amount of time?

   Ⓐ Lucy
   Ⓑ David
   Ⓒ They both took the same amount of time.
   Ⓓ There is not enough information given.

11. The Jamisons are on a road trip that will take 5 hours and 25 minutes. They have been driving for 3 hours and 41 minutes. How much longer do they need to travel before they reach their destination?

   Ⓐ 1 hour, 13 minutes
   Ⓑ 2 hours, 19 minutes
   Ⓒ 1 hour, 44 minutes
   Ⓓ 2 hours, 7 minutes

12. Rachel usually gets around 9 hours of sleep per night. She went to bed at 9:30 PM. About what time will she wake up?

   Ⓐ 8:30 AM
   Ⓑ 10:30 AM
   Ⓒ 6:30 AM
   Ⓓ 5:30 AM

13. A 45 minute long show ends at 12:20 PM. When did the show begin?

   Ⓐ 1:05 PM
   Ⓑ 11:35 AM
   Ⓒ 11:35 PM
   Ⓓ 11:45 AM

14. Mrs. James is giving her class a math test. She is allowing the students 40 minutes to finish the test. The test began at 10:22 AM. By what time must the test be finished?

   Ⓐ 10:42 AM
   Ⓑ 10:57 AM
   Ⓒ 11:02 AM
   Ⓓ 12:02 PM

15. The directions on a frozen pizza say to cook it for 25 minutes. Mr. Adams puts the frozen pizza in the oven at 5:43 PM. When will the pizza be done?

   Ⓐ 6:08 PM
   Ⓑ 6:18 PM
   Ⓒ 6:13 PM
   Ⓓ 5:58 PM

16. Which statements are true? Select all the correct answers.

Ⓐ The minute hand points to 4
Ⓑ The minute hand points to 6
Ⓒ The hour hand points to 6
Ⓓ The clock shows the time as 5:30

17. What time does this clock show? Write your answer in the box below.

18. Circle the clock that shows the time as 12:15

A   B   C

19. John starts working in the garden at 5:30 PM and finishes 40 minutes later. What time does the clock show when John finishes his work? Represent this on a number line.

20. The clocks in the first column show different times. For each clock in the first column, select the correct answer.

	9:42	11:58	2:03
Clock 1	○	○	○
Clock 2	○	○	○
Clock 3	○	○	○

Date of Completion:_____   Score:_____

CHAPTER 4 → Lesson 4: Elapsed Time

1. Cedric began reading his book at 9:12 AM. He finished at 10:02 AM. How long did it take him to read his book?

    Ⓐ 50 minutes
    Ⓑ 40 minutes
    Ⓒ 48 minutes
    Ⓓ 30 minutes

2. Samantha began eating her dinner at 7:11 PM and finished at 7:35 PM so that she could go to her room and play. How long did Samantha take to eat her dinner?

    Ⓐ 34 minutes
    Ⓑ 21 minutes
    Ⓒ 24 minutes
    Ⓓ 30 minutes

3. Tanya has after school tutoring from 3:00 PM until 3:25 PM. She began walking home at 3:31 PM and arrived at her house at 3:56 PM. How long did it take Tanya to walk home?

    Ⓐ 31 minutes
    Ⓑ 15 minutes
    Ⓒ 56 minutes
    Ⓓ 25 minutes

4. Doug loves to play video games. He started playing at 4:00 PM and did not finish until 5:27 PM. How long did Doug play video games?

    Ⓐ 1 hour and 37 minutes
    Ⓑ 1 hour and 27 minutes
    Ⓒ 27 minutes
    Ⓓ 2 hours and 27 minutes

5. Kelly has to clean her room before going to bed. She began cleaning her room at 6:12 PM. When she finished, it was 7:15 PM. How long did it take Kelly to clean her room?

    Ⓐ 57 minutes
    Ⓑ 53 minutes
    Ⓒ 1 hour and 3 minutes
    Ⓓ 1 hour and 15 minutes

6. Holly had a busy day. She attended a play from 7:06 PM until 8:13 PM. Then she went to dinner from 8:30 to 9:30 PM. How long did Holly attend the play?

   Ⓐ 57 minutes
   Ⓑ 2 hours and 27 minutes
   Ⓒ 46 minutes
   Ⓓ 1 hour and 7 minutes

7. Cara took her little brother to the park. They arrived at 3:11 PM and played until 4:37 PM. How long did Cara and her brother play at the park?

   Ⓐ 26 minutes
   Ⓑ 1 hour and 26 minutes
   Ⓒ 56 minutes
   Ⓓ 1 hour and 37 minutes

8. Arthur ran 5 miles. He began running at 8:19 AM and finished at 9:03 AM. How long did it take Arthur to run 5 miles?

   Ⓐ 44 minutes
   Ⓑ 45 minutes
   Ⓒ 40 minutes
   Ⓓ 54 minutes

9. Mr. Daniels wanted to see how fast he could wash the dishes. He began washing at 4:17 PM and finished at 4:32 PM. How long did it take Mr. Daniels to wash the dishes?

   Ⓐ 15 minutes
   Ⓑ 25 minutes
   Ⓒ 27 minutes
   Ⓓ 32 minutes

10. Sophia took a test that started at 3:28 PM. She finished the test at 4:11 PM. How long did it take Sophia to take her test?

    Ⓐ 37 minutes
    Ⓑ 47 minutes
    Ⓒ 33 minutes
    Ⓓ 43 minutes

11. Jonathan loves riding his bike, but he has to leave for football practice at 1:30 PM. If it is 1:11 PM now, how long does Jonathan have left to ride his bike before he has to leave for practice?

    Ⓐ 9 minutes
    Ⓑ 19 minutes
    Ⓒ 21 minutes
    Ⓓ 29 minutes

12. Mrs. Roberts loves to take a 20-minute nap on Saturdays. She was really tired when she went to sleep at 10:45 AM. She did not wake up until 11:25 AM. How long was Mrs. Roberts' long nap?

    Ⓐ 40 minutes
    Ⓑ 20 minutes
    Ⓒ 60 minutes
    Ⓓ 30 minutes

13. Spencer has to be at his piano lesson at noon. If it is now 11:29 AM, how long does Spencer have to get to his lesson?

    Ⓐ 31 minutes
    Ⓑ 1 minute
    Ⓒ 29 minutes
    Ⓓ 39 minutes

14. Look at the clocks below. How much time has elapsed between Clock A to Clock B?

    Clock A   Clock B

    Ⓐ 1 hour and 2 minutes
    Ⓑ 1 hour and 12 minutes
    Ⓒ 52 minutes
    Ⓓ 42 minutes

15. Look at the clocks below. How much time has elapsed between Clock A to Clock B?

Clock A   Clock B

Ⓐ 52 minutes
Ⓑ 42 minutes
Ⓒ 12 minutes
Ⓓ 32 minutes

16. Tiana's daily schedule consists of classes that are 45 minutes long. The table shows what time some of her classes start. What time will each class end? Select the correct answer.

	9:05	10:55	12:15
History 10:10	○	○	○
Math 8:20	○	○	○
Gym 11:30	○	○	○

17. Jonas is taking a long road trip. He drives for one hour and then stops and rests for 15 minutes. He repeats this until the end of the trip. Complete the table to show his schedule.

Driving Start Time	Break Time
1:00	2:00
2:15	
3:30	4:30

18. How much time has passed? Select all the correct answers.

Ⓐ 1 hour
Ⓑ 90 minutes
Ⓒ 2 hours
Ⓓ 120 minutes

19. Observe the two clocks. How many minutes have passed between the time shown in the first clock to the time in the second clock. Write your answer in the box given below.

20. Tim went out to do some work. He left home at 11:30 AM and returned back at 3:45 PM. How long was he away from home? Circle the correct answer.

Ⓐ 3 hours and 15 minutes
Ⓑ 4 hours and 15 minutes
Ⓒ 3 hours and 45 minutes
Ⓓ 4 hours and 45 minutes

CHAPTER 4 → Lesson 5: Solve word problems involving money

1. How much money in all is 1 dollar, 3 dimes, and 4 pennies?

    Ⓐ $1.19
    Ⓑ $1.79
    Ⓒ $1.30
    Ⓓ $1.34

2. What is the value of 2 quarters, 3 dimes, 4 nickels, and 1 penny?

    Ⓐ $1.01
    Ⓑ $2.01
    Ⓒ $0.91
    Ⓓ $1.41

3. What is the value of 3 dollars, 4 quarters, and 6 pennies?

    Ⓐ $4.46
    Ⓑ $3.46
    Ⓒ $4.06
    Ⓓ $3.64

4. Billy has 6 quarters, 3 dimes, and 2 nickels. How much money does Billy have in all?

    Ⓐ $1.80
    Ⓑ $1.90
    Ⓒ $2.80
    Ⓓ $2.90

5. Corey has $0.76. Which group of coins can Corey possibly have?

    Ⓐ Two quarters, six pennies
    Ⓑ Three quarters, one dime
    Ⓒ Five dimes, one quarter, one penny
    Ⓓ Seven dimes, six quarters

6. Breya has $2.13. Which group of coins can Breya possibly have?

    Ⓐ One dollar bill, four quarters, two nickels, three pennies
    Ⓑ Two dollar bills, one dime, thirteen pennies
    Ⓒ One dollar bill, four quarters, three pennies
    Ⓓ Two dollar bills, one nickel, three pennies

7. What is the value of 7 quarters, 5 nickels, and 25 pennies?

8. Match each combination of coins with the column of its correct total amount. All columns will not be selected.

	$1.47	$0.65	$1.57	$1.21	$1.65
4 quarters, 1 penny, 2 dimes					
2 dimes, 1 quarter, 10 pennies, 2 nickels					
9 dimes, 2 pennies, 1 nickel, 2 quarters					

9. Complete the table by filling in the value of the coins.

2 dollars, 1 quarter, 6 nickels, 1 dime	
5 dimes, 2 quarters, 3 nickels, 7 pennies	
3 quarters, 1 dollar, 3 dimes, 2 nickels, 20 pennies	

10. Bruce wants to buy a candy bar for $1.59. Bruce has 3 dimes, 3 quarters, and 11 nickels. Does he have enough money to buy the candy bar? Explain.

Date of Completion:_____   Score:_____

CHAPTER 4 → Lesson 6: Area

1. The area of a plane figure is measured in _____ units.

   Ⓐ cubic
   Ⓑ meter
   Ⓒ square
   Ⓓ box

2. Which of these objects has an area of about 1 square inch?

   Ⓐ a sheet of writing paper
   Ⓑ a beach towel
   Ⓒ a dollar bill
   Ⓓ a postage stamp

3. Mr. Parker wants to cover a mural with cloth. The mural is 12 inches long and 20 inches wide. How many square inches of cloth does Mr. Parker need?

   Ⓐ 240 square inches
   Ⓑ 32 square inches
   Ⓒ 120 square inches
   Ⓓ 220 square inches

4.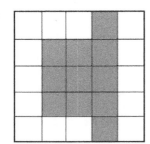

   What is the area of the shaded region?

   Ⓐ 10 square units
   Ⓑ 8 square units
   Ⓒ 11 square units
   Ⓓ 15 square units

5. Find the area of this figure.

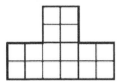

☐ =1 Square Unit

Ⓐ 22 square units
Ⓑ 20 square units
Ⓒ 18 square units
Ⓓ 16 square units

6. Find the area of this figure.

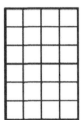

☐ =1 Square Unit

Ⓐ 22 square units
Ⓑ 20 square units
Ⓒ 24 square units
Ⓓ 28 square units

7. Find the area of the shaded region.

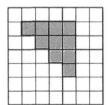

☐ =1 Square Unit

Ⓐ 11 square units
Ⓑ 10 square units
Ⓒ 16 square units
Ⓓ 9 square units

8. Find the area of the shaded region.

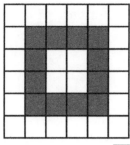

☐ = 1 Square Unit

Ⓐ 16 square units
Ⓑ 12 square units
Ⓒ 10 square units
Ⓓ 11 square units

9. Find the area of the shaded region.

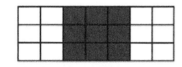

Key: ☐ = 1 Square Unit

Ⓐ 5 square units
Ⓑ 6 square units
Ⓒ 8 square units
Ⓓ 9 square units

10. The area of Karen's rectangular room is 72 sq. ft. If the length of the room is 8 ft. What is its width?

Ⓐ 8 ft.
Ⓑ 6 ft.
Ⓒ 7 ft.
Ⓓ 9 ft.

11. Can these items be measured in square units? Select yes or no.

	Yes	No
Window panes		
A ball		
Bathroom tile		
A banana		

12. Find the area of the shaded region in each figure. Each box is 1 square unit. Write your answers in the blank boxes in the table.

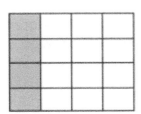 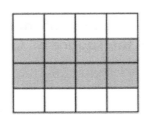 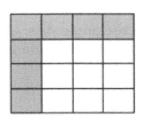

Figure A   Figure B   Figure C

Figure	Area
Figure A	
Figure B	
Figure C	

13. Find the area of the figure. Write your answer in the box given below.

6 feet

9 feet

 square feet

14. Which of the following are possible ways to find the area of this figure? Each box is 1 square unit. Select all correct answers.

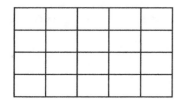

   Ⓐ Count the total number of square units
   Ⓑ Multiplying the length by the width of the figure
   Ⓒ Multiplying the number of square units by 2
   Ⓓ Subtracting the length of the figure from the width

15. The area of a rectangle A is 75 sq. cm. The area of square B is one third the area of the rectangle A. What is the side length of the square B? Circle the correct answer.

   Ⓐ 7 cm
   Ⓑ 5 cm
   Ⓒ 4 cm
   Ⓓ 6 cm

CHAPTER 4 → Lesson 7: Relating Area to Addition & Multiplication

1. Find the area of the rectangle below.

   3 feet ▬▬▬▬▬▬▬▬▬▬▬▬
              29 feet

   Ⓐ 87 square feet
   Ⓑ 32 square feet
   Ⓒ 64 square feet
   Ⓓ 128 square feet

2. Find the area of the rectangle below.

   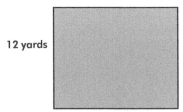

   12 yards
   15 yards

   Ⓐ 108 square yards
   Ⓑ 54 square yards
   Ⓒ 27 square yards
   Ⓓ 180 square yards

3. How could the area of this figure be calculated?

   33 inches
   63 inches

   Ⓐ Multiply 63 x 33 x 63 x 33
   Ⓑ Add 63 + 33 + 63 + 33
   Ⓒ Multiply 63 x 33
   Ⓓ Multiply 2 x 63 x 33

4. Find the area of the rectangle below.

Ⓐ 75 square meters
Ⓑ 50 square meters
Ⓒ 15 square meters
Ⓓ 30 square meters

5. Find the area of the rectangle below.

Ⓐ 176 square yards
Ⓑ 27 square yards
Ⓒ 54 square yards
Ⓓ 2,916 square yards

6. Find the area of the rectangle below.

Ⓐ 18 square inches
Ⓑ 15 square inches
Ⓒ 9 square inches
Ⓓ 6 square inches

7. Find the area of the rectangle below.

Ⓐ 16 square feet
Ⓑ 14 square feet
Ⓒ 9 square feet
Ⓓ 21 square feet

8. Find the area of the rectangle below.

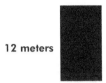

4 meters + 3 meters

Ⓐ 84 square meters
Ⓑ 48 square meters
Ⓒ 36 square meters
Ⓓ 72 square meters

9. Find the area of the rectangle below.

5 inches + 2 inches

Ⓐ 12 square inches
Ⓑ 10 square inches
Ⓒ 25 square inches
Ⓓ 35 square inches

10. Find the area of the rectangle below.

7 yards + 7 yards

Ⓐ 84 square yards
Ⓑ 182 square yards
Ⓒ 26 square yards
Ⓓ 19 square yards

11. The city wants to plant grass in a park. The park is 20 feet by 50 feet. How much grass will they need to cover the entire park?

Ⓐ 100 square feet
Ⓑ 500 square feet
Ⓒ 1,000 square feet
Ⓓ 1,100 square feet

12. Brenda wants to purchase a rug for her room. Her room is a rectangle that measures 7 yards by 6 yards. What is the area of her room?

    Ⓐ 42 square yards
    Ⓑ 48 square yards
    Ⓒ 36 square yards
    Ⓓ 26 square yards

13. Joan wants to cover her backyard with flowers. If her backyard is 30 feet long and 20 feet wide, what is the area that needs to be covered in flowers?

    Ⓐ 60 square feet
    Ⓑ 500 square feet
    Ⓒ 600 square feet
    Ⓓ 100 square feet

14. Bethany decided to paint the four walls in her room. If each wall measures 20 feet by 10 feet, how many total square feet will she need to paint?

    Ⓐ 400 square feet
    Ⓑ 200 square feet
    Ⓒ 800 square feet
    Ⓓ 600 square feet

15. Seth wants to cover his table top with a piece of fabric. His table is 2 meters long and 4 meters wide. How much fabric does Seth need?

    Ⓐ 6 square meters
    Ⓑ 10 square meters
    Ⓒ 8 square meters
    Ⓓ 16 square meters

CHAPTER 4 → Lesson 8: Perimeter

1. What is meant by the "perimeter" of a shape?

   Ⓐ The distance from the center of a plane figure to its edge
   Ⓑ The distance from one corner of a plane figure to an opposite corner
   Ⓒ The distance around the outside of a plane figure
   Ⓓ The amount of space covered by a plane figure

2. Complete the following statement.
   Two measurements associated with plane figures are _____.

   Ⓐ perimeter and volume
   Ⓑ perimeter and area
   Ⓒ volume and area
   Ⓓ weight and volume

3.

   This rectangle is 4 units long and one unit wide. What is its perimeter?

   Ⓐ 10 units
   Ⓑ 4 units
   Ⓒ 5 units
   Ⓓ 8 units

4.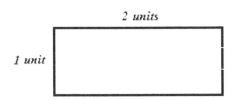

   What is the perimeter of the rectangle?

   Ⓐ 5 units
   Ⓑ 6 units
   Ⓒ 3 units
   Ⓓ 2 units

5.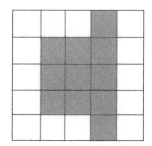

   What is the perimeter of the shaded region in the above figure?

   Ⓐ 16 units
   Ⓑ 15 units
   Ⓒ 11 units
   Ⓓ 10 units

6. The perimeter of this rhombus is 20 units. How long is each of its sides?

   Ⓐ 4 units
   Ⓑ 10 units
   Ⓒ 5 units
   Ⓓ This cannot be determined.

7. Each side of this rhombus measures 3 centimeters. What is its perimeter?

   Ⓐ 3 centimeters
   Ⓑ 12 centimeters
   Ⓒ 9 centimeters
   Ⓓ 6 centimeters

8. This square has a perimeter of 80 units. How long is each of its sides?

  Ⓐ 8 units
  Ⓑ 10 units
  Ⓒ 20 units
  Ⓓ 40 units

9. Find the perimeter of this figure.

   ☐ = 1 Square Unit

  Ⓐ 20 units
  Ⓑ 18 units
  Ⓒ 16 units
  Ⓓ 22 units

10. Joan wants to cover the outside border of her backyard with flowers. If her backyard is 30 feet long and 15 feet wide, how many feet of flowers does she need to plant?

  Ⓐ 450 feet
  Ⓑ 90 feet
  Ⓒ 60 feet
  Ⓓ 30 feet

11. Find the perimeter of this figure.

   ☐ =1 Square Unit

  Ⓐ 20 units
  Ⓑ 18 units
  Ⓒ 24 units
  Ⓓ 22 units

12. Brenda wants to place rope around a large field in order to play a game. The field is a rectangle that measures 23 yards by 32 yards. How much rope does Brenda need?

    Ⓐ 64 yards
    Ⓑ 736 yards
    Ⓒ 110 yards
    Ⓓ 55 yards

13. Find the perimeter of the following rectangle.

    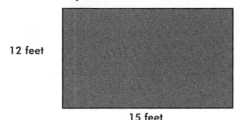

    Ⓐ 54 feet
    Ⓑ 27 feet
    Ⓒ 58 feet
    Ⓓ 180 feet

14. Find the perimeter of the following rectangle.

    Ⓐ 42 feet
    Ⓑ 176 feet
    Ⓒ 27 feet
    Ⓓ 54 feet

15. Find the perimeter of the shaded region.

    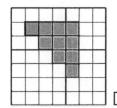

    = 1 Square Unit

    Ⓐ 10 units
    Ⓑ 13 units
    Ⓒ 15 units
    Ⓓ 16 units

16. Find the perimeter of the following rectangle.

Ⓐ 32 feet
Ⓑ 87 feet
Ⓒ 172 feet
Ⓓ 64 feet

17. The city is building a fence around a park. The park is 20 feet by 50 feet. How many feet of fencing do they need?

Ⓐ 100 feet
Ⓑ 120 feet
Ⓒ 140 feet
Ⓓ 70 feet

18. Find the perimeter of the following rectangle.

Ⓐ 86 inches
Ⓑ 172 inches
Ⓒ 1,749 inches
Ⓓ 50 inches

19. Find the perimeter of the following rectangle.

Ⓐ 15 meters
Ⓑ 25 meters
Ⓒ 30 meters
Ⓓ 50 meters

20. The perimeter of the following object is 16 feet. Find the length of the missing side.

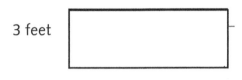

Ⓐ x = 5 feet
Ⓑ x = 10 feet
Ⓒ x = 13 feet
Ⓓ x = 6 feet

21. The perimeter of the following object is 20 feet. Find the length of the missing side.

Ⓐ x = 3 feet
Ⓑ x = 6 feet
Ⓒ x = 13 feet
Ⓓ x = 7 feet

22. The city is building a fence around a park. The park is 20 feet by 50 feet. If they only want the fence on 3 sides, what is the least amount of fencing they could buy?

Ⓐ 140 feet
Ⓑ 100 feet
Ⓒ 120 feet
Ⓓ 90 feet

23. The perimeter of the following object is 38 feet. Find the length of the missing side.

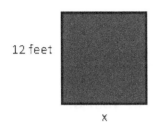

Ⓐ x = 17 feet
Ⓑ x = 12 feet
Ⓒ x = 7 feet
Ⓓ x = 26 feet

24. The perimeter of the following object is 24 inches. Find the length of the missing side.

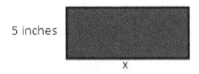

Ⓐ x = 19 inches
Ⓑ x = 7 inches
Ⓒ x = 9 inches
Ⓓ x = 12 inches

25. The perimeter of the following object is 54 yards. Find the length of the missing side.

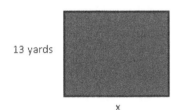

Ⓐ x = 28 yards
Ⓑ x = 14 yards
Ⓒ x = 41 yards
Ⓓ x = 27 yards

26. Which of the following statements are true? Select all correct answers.

   Ⓐ The perimeter is the distance around the outside of a plane figure.
   Ⓑ The perimeter is the center of a circle.
   Ⓒ The perimeter can be found by adding the angles of a triangle.
   Ⓓ The perimeter can be found by adding the length of a figure's sides.

27. What is the perimeter of the rectangle shown in figure below? Write your answer in the box given

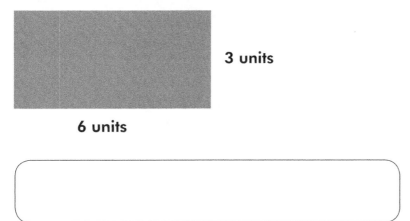

28. Circle the rhombus that has a perimeter of 8.0 cm.

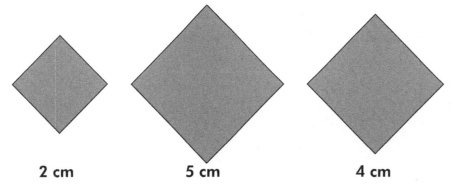

29. John draws a regular hexagon. Each side measures 8 centimeters. He also draws a regular octagon. Each side of the octagon measures 7 centimeters. Which shape has a greater perimeter? How did you arrive at the answer?

30. The perimeters of the rectangles are given in the first column. For each perimeter, select the possible areas of the rectangles.
Note that for each perimeter, more than one option may be correct.
Instruction: Assume that the length and the width of the rectangles are whole numbers.

	15 sq. cm.	10 sq. cm.	12 sq. cm.
Perimeter = 16 cm	☐	☐	☐
Perimeter = 14 cm	☐	☐	☐
Perimeter = 22 cm	☐	☐	☐

End of Measurement

ANSWER KEY AND DETAILED EXPLANATION

Chapter 4: Measurement

Lesson 1: Liquid Volume & Mass

Question No.	Answer	Detailed Explanation
1	D	The word "pound" after the number 40 indicates that 40 is a measurement of weight.
2	B	Kilograms are used to measure the mass of large, solid objects such as a table.
3	A	The amount of a liquid is also called its volume. Option C and D are not used to measure volume. Option B would be too large to measure water in a small bowl. Option A is the only logical choice.
4	B	Options A and C would not be used to measure weight. Option D would be too small to measure the weight of a table. Option B is the most logical choice.
5	A	Capacity is another word for volume. Options B, C, and D would not be used to measure volume. Option A is the most logical choice.
6	C	Option A is a measure for volume. Option B is a measure of distance or length. Option D is a measure of temperature. Option C is the most logical choice.
7	A	A gallon is a unit used to measure the volume of large amounts of liquid, whereas the other units measure smaller amounts of liquids.
8	C	Options A and B would be too light to be the weight of a typical 8-year-old, While Option D would be too heavy. Option C is the most logical choice.
9	A	Volume is a measurement associated with 3-dimensional figures. As a result, it is represented in cubic units. Volume is a measure of how many identical cubes (or parts of identical cubes) could fit within a solid figure.
10	B	Options C and D are not used to measure weight. Option A would be used to measure the weight of very large objects. Option B is the most logical choice.
11	C	Option D is not used to measure volume. Options A and B are both used to measure small amounts of liquids. Option C is the most logical choice for the large amount of water that takes up a swimming pool.

Question No.	Answer	Detailed Explanation
12	D	Options A, B, and C are all used to measure shorter lengths. Option D is the most logical choice.
13	A	Option C is not used to measure mass or volume. A cupcake recipe would have a very small amount of salt. Options B and D are both used to measure large quantities. Option A is the most logical choice for the small amount of salt that would be in Cupcakes.
14	A	Options B, C, and D are all used to measure shorter lengths and distances. Option A is used to measure large distances.
15	C	A millimeter is a very small unit of length (approximately the thickness of a fingernail).
16	Quarts & Liters	From the given options, quarts and liters are the only units used to measure liquids.
17	2 cups	The top measuring line of the cup is marked at 2 cups. This is the maximum amount of liquid it can hold. A measuring cup is a tool that can be used to measure quantities of liquid.
18	Teaspoon	From the given options, Teaspoon is the only tool used to measure small amounts of substances such as sugar.
19		This is a two-step problem. First, we calculate the total amount of water in the 8 coolers by multiplying the number of coolers (8) by the capacity of each water cooler (7 liters); 8 x 7 = 56 liters. Next, we calculate the amount of water consumed by subtracting the amount of water remained from the total amount of water; 56 - 5 = 51 liters.

Lesson 2: Measuring Length

Question No.	Answer	Detailed Explanation
1	C	Options A, B, and D are all customary units. Option C is the only metric unit.
2	B	A meter is just slightly longer than a yard. That is why a meterstick and a yardstick are almost the same length. 1 yard = 36 inches. 1 meter ≈ 39 inches.
3	C	Decimeters and centimeters are both too small to measure a table. Kilometers are used to measure long distances or lengths. Option C is the most appropriate.
4	C	Kilometers and meters are used to measure longer lengths. Grams are used to measure mass. Option C is the most appropriate.
5	A	Feet, yards, and miles are all used to measure longer lengths. The length of a pencil would be measured in inches.
6	A	Options B, C, and D are all too long to be the measure of a football. Option A is the most appropriate.
7	B	Option A and C are both too small for a football field. Option D is inappropriate because a football field is measurable. Option B is the most appropriate answer.
8	D	The width of a finger is small. Options A and B would both be too large. Option C is too small and would be more appropriate for the thickness of a fingernail. Option D is the most appropriate.
9	B	Hours are used to measure time. Yards are too small to measure a distance between two cities. Square inches are a measure of area, not distance. Option B is the most appropriate.
10	C	There are 2.54 centimeters in an inch. To convert centimeters to inches, divide 25 ÷ 2.54 = 9.84, rounding 9.84 to 10.
11	B	This ruler measures in inches. The end of the object stops at the number 8 so this object must be 8 inches.
12	C	This ruler measures in inches. The end of the object stops at the number 3 so this object must be 3 inches.
13	A	This ruler measures in inches. The end of the object stops at the number 6 so this object must be 6 inches.
14	B	Option A is false because inches are a smaller measurement unit than miles. Option C is false because a foot is longer than an inch. Option D is false because a mile is larger than a foot. Option B is the only statement that is true.
15	C	There are exactly 12 inches in one foot.

Question No.	Answer	Detailed Explanation
16	C & D	A ruler is a tool used to measure objects in inches. A ruler is usually no more than 12 inches long. From the answer choices, a carrot and a crayon can be measured with a ruler.
17	4 Inches	The length of the pencil is 4 inches according to the ruler. A ruler is a tool used to measure length.
18		

Measurement in inches	Measurement in half inches	Measurement in quarter inches
$2\frac{1}{2}$ inches	**5 half inches**	**10 quarter inches**
$5\frac{1}{2}$ inches	11 half inches	**22 quarter inches**
$6\frac{1}{2}$ inches	**13 half inches**	26 quarter inches

(1) 1 inch = 2 half inches; $2\frac{1}{2}$ inches = 2 inches + $\frac{1}{2}$ inch = 2 x 2 half inches + 1 half inch = 4 half inches + 1 half inch = 5 half inches.

1 half inch = 2 quarter inches. $2\frac{1}{2}$ inches = 5 half inches = 5 x 2 quarter inches = 10 quarter inches.

(2) 11 half inches = 10 half inches + 1 half inch = (10 ÷ 2) inches + $\frac{1}{2}$ inch = 5 inches + $\frac{1}{2}$ inch = $5\frac{1}{2}$ inches.

11 half inches = 11 x 2 quarter inches = 22 quarter inches.

(3) 26 quarter inches = 26 ÷ 2 half inches = 13 half inches.
13 half inches = 12 half inches + 1 half inch = (12 ÷ 2) inches + $\frac{1}{2}$ inch = 6 inches + $\frac{1}{2}$ inch = $6\frac{1}{2}$ inches.

Question No.	Answer	Detailed Explanation
19		

	4	6	8
How many fish are $16\frac{1}{2}$ inches long?	○	○	●
How many more fish are 16 inches long than 17 inches?	●	○	○
How many fish are less than $15\frac{3}{4}$ inches long?	○	●	○

X = 2 fish. There are 4 Xs at $16\frac{1}{2}$ inches. Therefore, the number of fish that are $16\frac{1}{2}$ inches long = 2 x 4 = 8

There are 4 Xs at 16 inches. Therefore, there are 8 (2 x 4) fish that are 16 inches long. There are 2 Xs at 17 inches. Therefore, there are 4 (2 x 2) fish which are 17 inches long. Therefore, there are 4 (8 - 4 = 4) more fish which are 16 inches long than 17 inches.

There are 2 (2 x 1) fish which are 15 inches long, and there are 4 (2 x 2) fish which are $15\frac{1}{2}$ inches long. Therefore, the number of fish which are less than $15\frac{3}{4}$ inches = 2 + 4 = 6.

Lesson 3: Telling Time

Question No.	Answer	Detailed Explanation
1	B	The hour hand (the shorter hand) is past the 2nd hour but has not reached the 3rd hour, and the minute hand (the longer hand) is past 15 minutes but has not yet reached 20 minutes.
2	D	The hour hand (the shorter hand) is past the 5th hour but has not reached the 6th hour, and the minute hand (the longer hand) is past 45 minutes but has not yet reached 50 minutes.
3	C	The hour hand (the shorter hand) is pointing to the 10th hour, and the minute hand (the longer hand) is past 0 minutes but has not yet reached 5 minutes.
4	A	The hour hand (the shorter hand) is past the 12th hour but has not reached the 1st hour, and the minute hand (the longer hand) is past 35 minutes but has not yet quite reached 40 minutes.
5	B	On a clock, the shorter hand points toward the hour and the longer hand points toward the minutes. For example, if it was 2:00, the shorter hand would point to the "2."
6	A	On a clock, the shorter hand points toward the hour and the longer hand points toward the minutes. For example, if it was 2:30, the longer hand would point to the "6," which represents the 30th minutes.
7	C	The hour hand (the shorter hand) is pointed at the 10th hour, and the minute hand (the longer hand) is at 2 minutes. The clock shows 10:02. Eight minutes after 10:02 would be 10:10.
8	A	The hour hand (the shorter hand) is past the 12th hour but not yet at the 1st hour, and the minute hand (the longer hand) is at 39 minutes. The clock shows 12:39. Twenty minutes after 12:39 would be 12:59.
9	B	The hour hand (the shorter hand) is past the 12th hour but not yet at the 1st hour, and the minute hand (the longer hand) is at 39 minutes. The clock shows 12:39. Ten minutes before 12:39 would be 12:29.
10	B	Lucy's time: 12:58 - 12:09 = 49 minutes. David's time: 1:03 - 12:15 = 48 minutes. David has the shorter time.

Question No.	Answer	Detailed Explanation
11	C	You can solve this problem by converting the hours to minutes and then subtracting the two times. 5 hours and 25 minutes is equivalent to (5 x 60) + 25 = 325 minutes. You multiply 5 hours by 60 because there are 60 minutes in an hour. 3 hours and 41 minutes is equivalent to (3 x 60) + 41 = 221 minutes. 325 - 221 = 104. Now convert 104 minutes back into hours and minutes by dividing by 60 and the answer is 1 hour and 44 minutes.
12	C	To calculate how many hours of sleep Rachel will receive, add the amount of time she sleeps to the time she goes to bed. 9 hours after 9:30 PM would be 6:30 AM.
13	B	To calculate when the show began, subtract the length of the show from the ending time. Counting back 45 minutes from 12:20 PM, you would arrive at 11:35 AM. The PM changes to AM, since you are now before noon.
14	C	To calculate when the students have to be finished with their test, add the amount of time given for the test to the start time. 40 minutes after 10:22 AM would be 11:02 AM.
15	A	To calculate when the pizza will be done, add the cooking time to the time Mr. Adams began cooking. 25 minutes after 5:43 PM would be 6:08 PM.
16	B & D	On an analog clock, the long hand shows the minutes while the short hand shows the hour. The minute hand on this clock points to 6 which represents 30 minutes. The hour hand is in between the numbers 5 and 6 which shows that the time is 5:30.
17	1:00	On an analog clock, the long hand shows the minutes while the short hand shows the hour. The minute hand on this clock points to 12 which represents an exact hour. The hour hand points to the number 1 shows that the time is exactly 1:00.
18	Clock A	Clock A is the correct answer. On an analog clock, the long hand shows the minutes while the short hand shows the hour. The minute hand on this clock points to 3 which represents 15 minutes. The hour hand is nearest to the number 12 which shows that the time is 12:15.

Question No.	Answer	Detailed Explanation
19		To determine what time John finishes his work, add 40 minutes to 5:30 PM; 5:30 PM + 40 minutes = 6:10 PM. This is represented on the number line below.

![number line from 5:30 to 6:15 with marks at 5:45, 6:00, 6:10]

In the figure, green dot shows the time when John started the work, and the red dot shows the time when John finished his work.

20

	9:42	11:58	2:03
Clock 1	○	●	○
Clock 2	○	○	●
Clock 3	●	○	○

(1) In the first clock, the hour hand (the shorter hand) has passed the 11th hour but not yet at the 12th hour.
At the start of the hour, the minute hand (the longer hand) points directly to 12, and it takes 5 minutes to move from one number to the next number and one minute to move from one tick to the next tick. So, the minute hand is at 58 minutes (5 × 11 + 3 = 58).
Therefore, the clock shows 11:58.

(2) In the second clock, the hour hand (the shorter hand) has passed the 2nd hour but not yet at the 3rd hour.
The minute hand is at 3 minutes (1 × 3 = 3).
Therefore, the clock shows 2:03.

(3) In the third clock, the hour hand (the shorter hand) has passed the 9th hour but not yet at the 10th hour.
The minute hand is at 42 minutes (5 × 8 + 2 = 42).
Therefore, the clock shows 9:42.

Lesson 4: Elapsed Time

Question No.	Answer	Detailed Explanation
1	A	Counting back from 10:02 to 10:00 is 2 minutes. Then, counting back from 10:00 back to 9:12 is an additional 48 minutes, making the total elapsed time 50 minutes.
2	C	Subtract the beginning time from the ending time; 7:35 back to 7:11 is 24 minutes.
3	D	Subtract the time Tanya began walking from the time she arrived home; 3:56 back to 3:31 is 25 minutes.
4	B	From 4:00 to 5:00 is one hour of elapsed time. From 5:00 until 5:27 is an additional 27 minutes, for a total elapsed time of 1 hour and 27 minutes.
5	C	From 6:12 to 7:12 is 1 hour of elapsed time. From 7:12 to 7:15 is an additional 3 minutes, for a total elapsed time of 1 hour and 3 minutes.
6	D	From 7:06 to 8:06 is one hour of elapsed time. From 8:06 to 8:13 is an additional 7 minutes, for a total elapsed time of 1 hour and 7 minutes.
7	B	From 3:11 to 4:11 is one hour of elapsed time. From 4:11 to 4:37 is an additional 26 minutes, for a total elapsed time of 1 hour and 26 minutes.
8	A	Counting back from 9:03 to 9:00 is 3 minutes. Then, counting back from 9:00 to 8:19 is an additional 41 minutes, for a total elapsed time of 44 minutes.
9	A	Subtract the beginning time from the ending time; 4:32 back to 4:17 is 15 minutes.
10	D	Counting back from 4:11 to 4:00 is 11 minutes. Then, counting back from 4:00 to 3:28 is an additional 32 minutes, for a total elapsed time of 43 minutes.
11	B	Subtract the present time from the time he has to leave; 1:30 - 1:11 = 19 minutes
12	A	From 10:45 to 11:00 is 15 minutes of elapsed time. From 11:00 to 11:25 is an additional 25 minutes, for a total elapsed time of 40 minutes.
13	A	Counting back from noon (12:00) to 11:30 is 30 minutes. Then, counting back from 11:30 to 11:29 is an additional minute, for a total elapsed time of 31 minutes.
14	A	Clock A shows 12:01 and Clock B shows 1:03. The elapsed time from 12:01 to 1:03 is 1 hour and 2 minutes.
15	D	Clock A shows 7:15 and Clock B shows 7:47. The elapsed time from 7:15 to 7:47 is 32 minutes.

Question No.	Answer	Detailed Explanation
16		

	9:05	**10:55**	**12:15**
History 10:10	○	●	○
Math 8:20	●	○	○
Gym 11:30	○	○	●

In order to find the end time, add 45 minutes to the start time. Time is measured in 60 minute intervals. If the total exceeds 60 then add time to the next hour. To add 45 minutes to 8:20, add 40 minutes which will reach 9:00. Then add the remaining 5 minutes. The end time will be 9:05. Using the same strategy, 10:10 plus 45 minutes will be 10:55 and 11:30 plus 45 minutes will be 12:15.

Question No.	Answer	Detailed Explanation
17		

Driving Start Time	Break Time
1:00	2:00
2:15	**3:15**
3:30	4:30
4:45	**5:45**

The correct answers are 3:15, 4:45 and 5:45. To find the break time, add one hour to the start time. Time is measured in 60 minute intervals. If the total exceeds 60 then add time to the next hour. To find the next start time after the break, add 15 minutes to the break time.

Question No.	Answer	Detailed Explanation
18	C & D	Time is measured in 60 minute intervals. 60 minutes is 1 hour. The time is 4 o'clock in the first picture. The time is 6 o'clock in the second picture. 2 hours have elapsed. 2 hours can also be seen as 120 minutes since 60+60=120
19	30 minutes	The time is 1 o'clock in the first picture. The time is 1:30 the second picture. 30 minutes have elapsed.
20	B	We have to subtract the time Tim left home from the time he returned; 3:45 PM - 11:30 AM. From 11:30 AM to 12 noon is 30 minutes of elapsed time. From 12 noon to 3:00 PM is 3 hours of elapsed time. From 3:00 PM to 3:45 PM is 45 minutes of elapsed time. Therefore, total elapsed time = 30 minutes + 3 hours + 45 minutes = 3 hours and 75 minutes. 3 hours and 75 minutes = 4 hours and 15 minutes.

Lesson 5: Solve Word Problems Involving Money

Question No.	Answer	Detailed Explanations
1	D	The answer is D. The value of a dime is 10 cent, so three dimes is 30 cents. The value of a penny is 1 cent, so 4 pennies is 4 cents. If you add them together, you get the total of $1.34.
2	A	The answer is A. The value of 2 quarters is 0.50, the value of 3 dimes is 0.30, the value of 4 nickels is 0.20, and the value of 1 penny is 0.01. If you add them together, you get the total of $1.01.
3	C	The answer is C. 3 dollars have a value of 3.00, 4 quarters have a value of 1.00, and 6 pennies have a value of 0.06. If you add them together you get $4.06.
4	B	The answer is B. 6 quarters have a value of 1.50, 3 dimes have a value of 0.30, and 2 nickels have a value of 0.10. If you add them together, you get $1.90.
5	C	The answer is C. Five dimes have a value of 0.50. One quarter have a value of 0.25 and one penny have a value of 0.01. If you add them together, you get $0.76.
6	A	The answer is A. The value of one dollar bill is 1.00, the value of four quarters is 1.00, the value of two nickels is 0.10, and the value of three pennies is 0.03. If you add them together you get $2.13.
7	$ 2.25	The answer is $ 2.25.
8		4 quarters, 1 penny, 2 dimes --> **$1.21** 2 dimes, 1 quarter, 10 pennies, 2 nickels --> **$0.65** 9 dimes, 2 pennies, 1 nickel, 2 quarters --> **$1.47**
9		2 dollars, 1 quarter, 6 nickels, 1 dime --> **$2.65** 5 dimes, 2 quarters, 3 nickels, 7 pennies --> **$1.22** 3 quarters, 1 dollar, 3 dimes, 2 nickels, 20 pennies --> **$2.35**
10		Bruce have enough money to buy the candy bar. If Bruce has 30 cent, 75 cent, and 55 cent, then he has the total amount of $1.60.

Lesson 6: Area

Question No.	Answer	Detailed Explanation
1	C	Area is a 2-dimensional attribute, so it must be represented in square units. Area is a measure of how many identical squares (or parts of identical squares) would be needed to cover a figure.
2	D	A postage stamp is a rectangle measuring about 1 inch on each side. Therefore, the area of a postage stamp is about 1 square inch. The other objects are all too large to measure 1 square inch as this is a very small measurement.
3	A	Area of a rectangle is calculated by multiplying length by width: 12 inches x 20 inches = 240 square inches.
4	C	If each box is a square unit, count the number of shaded boxes to get the area of the shaded region. There are 11 shaded boxes so the area is equal to 11 square units.
5	D	If each box is a square unit, count the number of boxes to get the area. There are 16 boxes so the area is equal to 16 square units.
6	C	If each box is a square unit, count the number of boxes to get the area. There are 24 boxes so the area is equal to 24 square units.
7	B	If each box is a square unit, count the number of shaded boxes to get the area of the shaded region. There are 10 shaded boxes so the area is equal to 10 square units.
8	B	If each box is a square unit, count the number of shaded boxes to get the area of the shaded region. There are 12 shaded boxes so the area is equal to 12 square units.
9	D	If each box is a square unit, count the number of shaded boxes to get the area of the shaded region. There are 9 shaded boxes so the area is equal to 9 square units.
10	D	Area of a rectangle = length x width. In this problem, we know the area and the length of the room. Let the width be w. 72 = 8 x w. What is the number when multiplied by 8 gives 72? It is 9. Therefore, w = 9 ft.

Question No.	Answer	Detailed Explanation
11		

	Yes	No
Window panes	✓	
A ball		✓
Bathroom tile	✓	
A banana		✓

Question No.	Answer	Detailed Explanation
12	4; 8; 7	The correct answers are 4, 8, and 7. Each shaded box is one square unit. Figure A has 4 square units shaded. Figure B has 8 square units shaded. Figure C has 7 square units shaded.
13	54 square feet	To find the area of a rectangle, multiply length by width. 6 feet x 9 feet = 54 square feet.
14	A & B	Each box is one square unit. Count the total number of square units in order to find the area. Another method is to count the square units along the width and multiply the total by the number of square units along the length.
15	B	Area of the square B is one third the area of the rectangle A. It means, we have to divide the area of the rectangle A by 3 to get the area of the square B. Area of the square B = 75 ÷ 3 = 25 sq. cm. Area of a square = side length x side length. Area of the square B = 25 sq. cm. What is the number when multiplied by itself will give 25? It is 5. Therefore, side length = 5 cm.

Lesson 7: Relating Area to Addition & Multiplication

Question No.	Answer	Detailed Explanation
1	A	Area is calculated by multiplying length by width: 3 feet x 29 feet = 87 square feet. (Note: 3 x 29 = 29 + 29 + 29 = 87)
2	D	Area is calculated by multiplying length by width: 12 yards x 15 yards = 180 square yards. [Note: 15 x 12 = (15 x 10) + (15 x 2) = 150 + 30 = 180]
3	C	Area of a rectangle is calculated by multiplying length by width. To find the area of this rectangle, multiply 63 x 33.
4	B	Area of a rectangle is calculated by multiplying length by width: 5 meters x 10 meters = 50 square meters.
5	A	Area of a rectangle is calculated by multiplying length by width: 16 yards x 11 yards = 176 square yards. [Note: 16 x 11 = (16 x 10) + (16 x 1) = 160 + 16 = 176]
6	B	Area of a rectangle is calculated by multiplying length by width: 3 inches x (2 + 3) inches = 3 inches x 5 inches = 15 square inches.
7	D	Area of a rectangle is calculated by multiplying length by width: 7 feet x (2 + 1) feet = 7 feet x 3 feet = 21 square feet.
8	A	Area of a rectangle is calculated by multiplying length by width: 12 meters x (4 + 3) meters = 12 meters x 7 meters = 84 square meters.
9	D	Area of a rectangle is calculated by multiplying length by width: 5 inches x (5 + 2) inches = 5 inches x 7 inches = 35 square inches.
10	B	Area of a rectangle is calculated by multiplying length by width: 13 yards x (7 x 7) yards = 13 yards x 14 yards = 182 square yards.
11	C	The park is rectangular. The area of a rectangle is calculated by multiplying length by width: 50 feet x 20 feet = 1,000 square feet.
12	A	Area of a rectangle is calculated by multiplying length by width: 7 yards x 6 yards = 42 square yards.
13	C	Area of a rectangle is calculated by multiplying length by width: 30 feet x 20 feet = 600 square feet.
14	C	The area of one wall is calculated by multiplying length by width: 20 feet x 10 feet = 200 square feet. There are four identical walls, so multiply 4 by 200 to calculate the total area: 4 x 200 square feet = 800 square feet.
15	C	Area is calculated by multiplying length by width: 2 meters x 4 meters = 8 square meters.

Lesson 8: Perimeter

Question No.	Answer	Detailed Explanation
1	C	The perimeter of a shape is the distance around the shape.
2	B	A plane figure is a two-dimensional (flat) shape. Area and perimeter are both associated with these types of objects. Volume and weight apply to 3-dimensional figures.
3	A	To find the perimeter of a rectangle, total the lengths of its four sides. 4 + 1 + 4 + 1 = 10 units
4	B	To find the perimeter of a rectangle, total the lengths of its four sides. 1 + 2 + 1 + 2 = 6 units
5	A	To find the perimeter of the shaded area, count the edges that surround the outside of the shaded area. There are 16 sides around the shaded area so the perimeter is 16.
6	C	A rhombus has 4 equal sides and the perimeter is equal to the sum of those 4 sides. To calculate the length of each side, divide the perimeter by 4. 20 ÷ 4 = 5.
7	B	Because a rhombus has four equal sides, to find its perimeter multiply the length of one of its sides by four. 3 x 4 = 12 centimeters
8	C	A square has 4 equal sides and the perimeter is equal to the sum of those 4 sides. To calculate the length of each side, divide the perimeter by 4. 80 ÷ 4 = 20.
9	A	To find the perimeter of the figure, count the edges that surround the outside of the figure. There are 20 sides around the figure so the perimeter is 20.
10	B	The yard is rectangular. To find its perimeter, total the lengths of its four sides. 30 + 15 + 30 + 15 = 90 feet
11	A	To find the perimeter of the figure, count the edges that surround the outside of the figure. There are 20 sides around the figure so the perimeter is 20.
12	C	The field is rectangular. To find its perimeter, add the lengths of its four sides. 23 + 32 + 23 + 32 = 110 yards
13	A	To find the perimeter of a rectangle, total the lengths of its four sides. 15 + 12 + 15 + 12 = 54 feet
14	D	To find the perimeter of a rectangle, total the lengths of its four sides. 16 + 11 + 16 + 11 = 54 feet

Question No.	Answer	Detailed Explanation
15	D	To find the perimeter of the shaded area, count the edges that surround the outside of the shaded area. There are 16 sides around the shaded area so the perimeter is 16.
16	D	To find the perimeter of a rectangle, total the lengths of its four sides. 3 + 29 + 3 + 29 = 64 feet
17	C	The park is rectangular. To find its perimeter, total the lengths of its four sides. 20 + 50 + 20 + 50 = 140 feet
18	B	To find the perimeter of a rectangle, total the lengths of its four sides. 33 + 53 + 33 + 53 = 172 inches
19	C	To find the perimeter of a rectangle, total the lengths of its four sides. 10 + 5 + 10 + 5 = 30 meters
20	A	To solve for the missing side, plug in what you know into the formula for perimeter: 2W + 2L = P (2 x 3) + 2x = 16 To solve for x: 2x = 16 - 6 = 10. Divide 10 by 2 to find one side; x = 5 feet.
21	A	To solve for the missing side, plug in what you know into the formula for perimeter: 2W + 2L = P (2 x 7) + 2x = 20 To solve for x: 2x = 20 - 14 = 6 Divide 6 by 2 to find one side; x = 3 feet.
22	D	If there are 4 sides to the rectangular park, the measurements of each side would be 20, 50, 20, 50. The question asks for the least amount of fencing needed for 3 sides. They should choose to fence the two 20-foot sides and one of the 50-foot sides. The amount of fencing material needed would be 20 + 20 + 50 = 90 feet.
23	C	To solve for the missing side, plug in what you know into the formula for perimeter: 2W + 2L = P. (2 x 12) + 2x = 38, now solve for x. To solve for x: x = (38 - 24) ÷ 2, x = 7.
24	B	To solve for the missing side, plug in what you know into the formula for perimeter: 2W + 2L = P (2 x 5) + 2x = 24 To solve for x : 2x = 24 - 10 = 14 Divide 14 by 2 to find one side; x = 7 inches.

LumosLearning.com

Question No.	Answer	Detailed Explanation
25	B	To solve for the missing side, plug in what you know into the formula for perimeter: 2W + 2L = P (2 x 13) + 2x = 54 To solve for x: 2x = 54 - 26 = 28 Divide 28 by 2 to find one side; x = 14 yards.
26	A & D	The perimeter is the distance around the outside of a plane figure. The perimeter can be found by adding the length of a figure's sides.
27	18 units	The perimeter can be found by adding the length of a figure's sides. 3 + 3 + 6 + 6 = 18.
28	2 cm	A rhombus has 4 equal sides. Perimeter can be found by adding the length of a figure's sides. Perimeter of the figure is 8 cm. Rhombus with four 2 cm sides is the only figure that has a perimeter of 8 cm., because 2 + 2 + 2 + 2 = 8 cm.
29		A regular hexagon has six equal sides. Therefore, perimeter of the hexagon = 6 x length of one side = 6 x 8 = 48 cm. A regular octagon has eight equal sides. Therefore, perimeter of the octagon = 8 x length of one side = 8 x 7 = 56 cm. 56 cm > 48 cm. Therefore, perimeter of the octagon > perimeter of the hexagon

Question No.	Answer	Detailed Explanation

30

	15 sq. cm.	10 sq. cm.	12 sq. cm.
Perimeter = 16 cm	✓		✓
Perimeter = 14 cm		✓	✓
Perimeter = 22 cm		✓	

Let L be the length of the rectangle and W be the width of the rectangle. Perimeter = 2 (L + W)

(1) Perimeter = 16; 2 x (L + W) = 16; L + W = 16 ÷ 2 = 8. We have to find the length and the width of the rectangle whose sum is 8 cm, and then find the area. Find the areas of the rectangles by substituting L = 1, 2, 3 etc. Among the choices given, there are 2 possible areas. (a) If L = 2, W = 8 - 2 = 6 . Area = L x W = 2 x 6 = 12 sq. cm. (b) If L = 3, W = 8 - 3 = 5. Area = 3 x 5 = 15 sq. cm.

(2) Perimeter = 14; 2 x (L + W) = 14; L + W = 14 ÷ 2 = 7. We have to find the length and the width of the rectangle whose sum is 7 cm, and then find the area. Find the areas of the rectangles by substituting L = 1, 2, 3 etc. Among the choices given, there are 2 possible areas. (a) If L = 2, W = 7 - 2 = 5. Area = 2 x 5 = 10 sq. cm. (b) If L = 3, W = 7 - 3 = 4. Area = 3 x 4 = 12 sq. cm.

(3) Perimeter = 22; 2 x (L + W) = 22; L + W = 22 ÷ 2 = 11. We have to find the length and the width of the rectangle whose sum is 11 cm, and then find the area. Find the areas of the rectangles by substituting L = 1, 2, 3 etc. Among the choices given, there is only one possibility. If L = 1, W = 11 - 1 = 10. Area = 1 x 10 = 10 sq. cm.

Date of Completion:_____ Score:_____

Chapter 5: Data Analysis

Lesson 1: Graphs

1.

Class Survey Should there be a field trip?		
	Yes	No
Mr. A's class	|||| |||| ||||	|||| ||
Mrs. B's class	|||| |||| ||||	|||| |||| |||
Mr. C's class	|||| ||||		|||| ||||	
Mrs. D's class	|||| |||| ||	|||| ||||

Four 3rd grade classes in Hill Elementary School were surveyed to find out if they wanted to go on a field trip at the end of the school year. The tally table above was used to record the votes.
How many kids voted "Yes" in Mrs. B's class?

Ⓐ 28 kids
Ⓑ 15 kids
Ⓒ 13 kids
Ⓓ 23 kids

2.

Should there be a field trip?		
	Yes	No
Mr. A's class	14	7
Mrs. B's class	15	13
Mr. C's class	11	11
Mrs. D's class	12	10
Total	52	41

Four 3rd grade classes in Hill Elementary School were surveyed to find out if they wanted to go on a field trip at the end of the school year. The table above shows the results of the survey.
How many kids voted "Yes" in Mr. A's class?

Ⓐ 7 kids
Ⓑ 15 kids
Ⓒ 14 kids
Ⓓ 21 kids

3.

Should there be a field trip?		
	Yes	No
Mr. A's class	14	7
Mrs. B's class	15	13
Mr. C's class	11	11
Mrs. D's class	12	10
Total	52	41

Four 3rd grade classes in Hill Elementary School were surveyed to find out if they wanted to go on a field trip at the end of the school year. The table above shows the results of the survey.
How many kids altogether voted "No" for the field trip?

Ⓐ 82 kids
Ⓑ 11 kids
Ⓒ 52 kids
Ⓓ 41 kids

4. The students in Mr. Donovan's class were surveyed to find out their favorite school subjects. The results are shown in the pictograph. Use the pictograph to answer the following question:
How many students chose either science or math?

Our Favorite Subjects

Math	○○○○
Reading	○○
Science	○○○
History	○
Other	○○

Key: ○ = 2 votes

Ⓐ 6 students
Ⓑ 7 students
Ⓒ 14 students
Ⓓ 2 students

5.

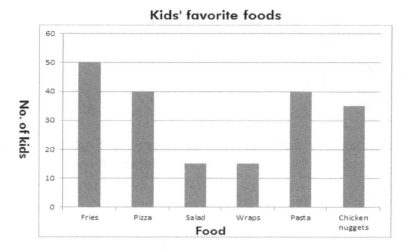

The third graders in Valley Elementary School were asked to pick their favorite food from 6 choices. The results are shown in the bar graph.
Which food was the favorite of the most children?

Ⓐ Pizza
Ⓑ Pasta
Ⓒ Fries
Ⓓ Salad

6.

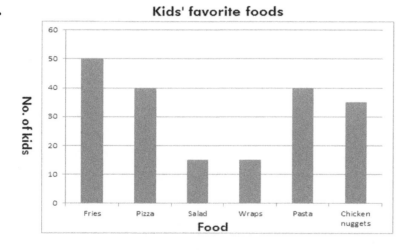

The third graders in Valley Elementary School were asked to pick their favorite food from 6 choices. The results are shown in the bar graph.

What are the 2 foods that kids like the least?

Ⓐ Fries and Pizza
Ⓑ Pizza and Pasta
Ⓒ Pasta and Chicken Nuggets
Ⓓ Salad and Wraps

7.

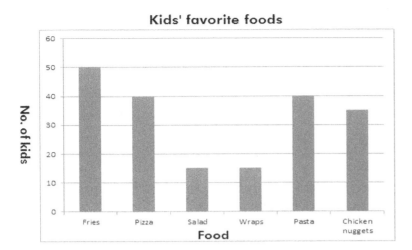

The third graders in Valley Elementary School were asked to pick their favorite food from 6 choices. The results are shown in the bar graph.
How many kids chose pasta?

Ⓐ 50 kids
Ⓑ 15 kids
Ⓒ 40 kids
Ⓓ 35 kids

8.

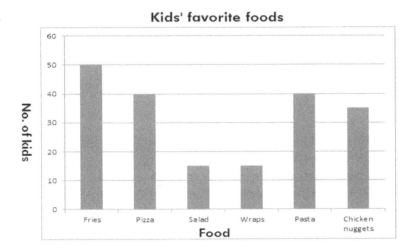

The third graders in Valley Elementary School were asked to pick their favorite food from 6 choices. The results are shown in the bar graph.

How many more kids prefer fries than pizza?

Ⓐ 50 more kids
Ⓑ 10 more kids
Ⓒ 1 more kid
Ⓓ 15 more kids

9.

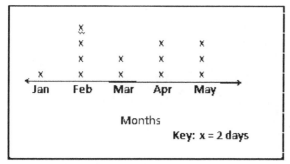

The line plot shows the number of days it rained in New Jersey from January through May. What should be the title of the above graph?

Ⓐ Line plot
Ⓑ Rainy Days in New Jersey
Ⓒ Months
Ⓓ 2 days

10.

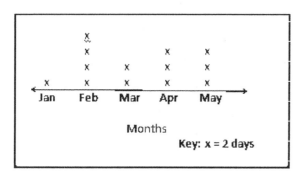

Which of the following statements about the above graph is true?

Ⓐ The graph shows New Jersey's monthly rainy days from January through May.
Ⓑ The graph shows the amount of rain accumulated each day.
Ⓒ The graph shows the average temperature during the 5 month period.
Ⓓ The graph shows New Jersey's total number of rainy days for the year.

11.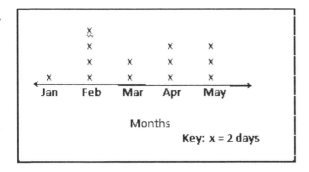

According to the graph, which month had the most rainy days?

Ⓐ March
Ⓑ February
Ⓒ January
Ⓓ April

12. A survey was taken to find out the favorite sports of third graders in a particular class. The results are shown in the tally table. Use the table to answer the following question:
How many students were surveyed altogether?

Our Favorite Sports

Soccer	𝍤 l
Tennis	llll
Baseball	𝍤 lll
Hockey	llll
Other	lll

Ⓐ 20 students
Ⓑ 25 students
Ⓒ 24 students
Ⓓ 27 students

13. A survey was taken to find out the favorite sports of third graders in a particular class. The results are shown in the tally table. Use the table to answer the following question: How many more students chose soccer than chose hockey?

Our Favorite Sports

Soccer	||||	
Tennis	||||	
Baseball	|||| |||
Hockey	||||	
Other	|||	

Ⓐ 6 students
Ⓑ 4 students
Ⓒ 2 students
Ⓓ 3 students

14. A survey was taken to find out the favorite sports of third graders in a particular class. The results are shown in the tally table. Use the table to answer the following question: How many students chose baseball as their favorite sport?

Our Favorite Sports

Soccer	||||	
Tennis	||||	
Baseball	|||| |||
Hockey	||||	
Other	|||	

Ⓐ 9 students
Ⓑ 8 students
Ⓒ 3 students
Ⓓ 6 students

15. A survey was taken to find out the favorite sports of third graders in a particular class. The results are shown in the tally table. Use the table to answer the following question: Which two sports were chosen by the same number of students?

Our Favorite Sports

Soccer	𝄁𝄁𝄁𝄁	
Tennis	||||	
Baseball	𝄁𝄁𝄁𝄁	|||
Hockey	||||	
Other	|||	

Ⓐ soccer and tennis
Ⓑ soccer and baseball
Ⓒ hockey and soccer
Ⓓ hockey and tennis

16. Mrs. Brown's class voted on which day they will have a class party. Look at the graph. Each figure represents 1 student. Match the correct answers to the number of votes.

Student Party Day Votes

	5	2	4
Total votes for Friday	○	○	○
Total votes for Wednesday	○	○	○
Total votes for Monday	○	○	○

17. Coach Dennis is creating a graph. He wants to purchase bats for his team. He needs to purchase 8 bats. He needs half of the bats to be made of wood. The other half will be made of aluminum. He decides that he will purchase 3 more bats made of plastic as well. Complete the table by filling in the correct answers.

Type of Bat	Number of Bats Needed
Wood	
Aluminum	
Plastic	

18. Vicki is planting a flower garden. The graph above shows the number of flowers to be planted in the garden. Which of the following statements are true? Select all the correct answers.

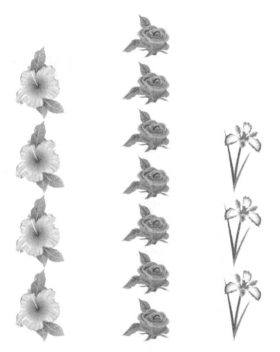

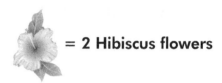 = 2 Hibiscus flowers

 = 2 Rose flowers

 = 2 Iris flowers

Ⓐ The count of Rose flowers is 14
Ⓑ The Count of Rose flowers is 7
Ⓒ The count of Hibiscus flowers is 4
Ⓓ The count of Iris flowers is 6

19. Find the total number of each coin. Use the tally chart to draw a bar graph.

Coins in John's Piggy Bank		
Coin	Tally	Number of Coins
Penny	|||| |||| |||| |||| ||	
Nickel	|||| |||| |||| |||	
Dime	|||| |||| ||||	
Quarter	|||| |||| ||||		

20. From the Venn Diagram given below, represent the number of people who only own cats as pet to the number of people who own only dogs as a pet in the form of a fraction $\frac{a}{b}$ (ratio of number of people owning cats to dogs)

ANSWER KEY AND DETAILED EXPLANATION

Chapter 5: Data Analysis

Lesson 1: Graphs

Question No.	Answer	Detailed Explanation
1	B	The chart shows 3 sets of 5 tallies for Mrs. B's class in the "yes" column. Multiplying 3 x 5 the tallies represent 15 kids.
2	C	The number 14 in the "yes" column for Mrs. A's class represents 14 votes.
3	D	The "Total" row displays the overall number of votes. There is a total of 41 votes represented in the "No" column.
4	C	First, add the totals number of students who chose Science and Math. 3 + 4 = 7. The chart states that each object stands for 2 votes. Multiply the Science and Math total by 2. 7 x 2 = 14.
5	C	The tallest bar indicates the food that was chosen most often. That would be considered the "favorite."
6	D	The foods with the shortest bars represent the foods that were least liked by the kids. Both salads and wraps had the least amount of votes.
7	C	Locate "pasta" at the bottom of the graph. The bar for pasta reaches up to the 40 line.
8	B	First find the values for both fries and pizza by locating them on the x-axis and then moving over to the y-axis to see their value. Subtract the number of kids who chose pizza from the number of kids who chose fries. 50 - 40 = 10.
9	B	The title of the graph is located above the graph.
10	A	Option B is false because the graph makes no mention of amount of rain. Option C is false because the title of the graph states "rainy days" and not temperature. Option D is false because the graph only shows 5 months which is not equivalent to a year. Option A is the only choice that is true.

Question No.	Answer	Detailed Explanation
11	B	According to the graph, January had 2 rainy days, February had 8 rainy days, March had 4 rainy days, and April had 6 rainy days. February had the most rainy days.
12	B	Add up all the tallies to obtain the total. 6 + 4 + 8 + 4 + 3 = 25.
13	C	Subtract the number of students who chose hockey from the number of students who chose soccer. 6 - 4 = 2.
14	B	There are 8 tally marks in the baseball section. This represents the 8 students who voted for baseball as their favorite sport.
15	D	Hockey and tennis both have 4 tallies on the chart.
16		

	5	2	4
Total votes for Friday	●	○	○
Total votes for Wednesday	○	○	●
Total votes for Monday	○	●	○

Each figure in the graph represents 1 student. 5 students voted for Friday, 4 students voted for Wednesday, and 2 students voted for Monday.

17		

Type of Bat	Number of Bats Needed
Wood	4
Aluminum	4
Plastic	3

Coach Dennis needs 4 wooden bats, 4 aluminum bats, and 3 plastic bats. "Half of the bats" means that 8 bats will need to be divided in half or by 2. 8 divided by 2 equals four.

| 18 | A & D | Each flower on the graph represents 2 flowers. There are 8 Hibiscus flowers, 14 Rose flowers, and 6 Iris flowers. |

Question No.	Answer	Detailed Explanation
19		Every fifth mark is drawn across the previous four marks.

In the tally of penny coins, there are four 5s (making it 4 x 5 = 20) and two singles (2 x 1 = 2). So, there are 22 pennies. Similarly, we can find that there are 18 nickels, 14 dimes and 16 quarters.

Coins in John's Piggy Bank		
Coin	Tally	Number of Coins
Penny	|||| |||| |||| |||| ||	22
Nickel	|||| |||| |||| |||	18
Dime	|||| |||| ||||	14
Quarter	|||| |||| ||||		16

Tally chart is used to draw the bar graph given below.

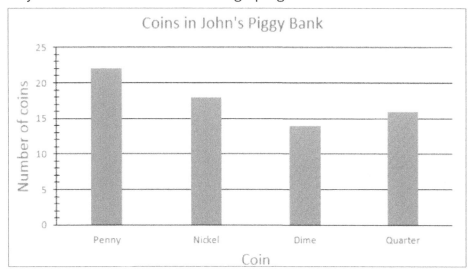

Question No.	Answer	Detailed Explanation
20		The left side of the Venn diagram shows the number of people who own only cats. This is 2 people. The right side shows the number of people who own only dogs which is 3 people. The center overlapping portion shows the number of people who own both cat and dog which is 2. We need number of people who own only cats / the number of people who own only dogs which is $\frac{2}{3}$.

Additional Information

Test Taking Tips

1) **The day before the test,** make sure you get a good night's sleep.

2) **On the day of the test,** be sure to eat a good hearty breakfast! Also, be sure to arrive at school on time.

3) During the test:

- **Read every question carefully.**

  - Do not spend too much time on any one question. Work steadily through all questions in the section.
  - Attempt all of the questions even if you are not sure of some answers.
  - If you run into a difficult question, eliminate as many choices as you can and then pick the best one from the remaining choices. Intelligent guessing will help you increase your score.
  - Also, mark the question so that if you have extra time, you can return to it after you reach the end of the section.
  - Some questions may refer to a graph, chart, or other kind of picture. Carefully review the graphic before answering the question.
  - Be sure to include explanations for your written responses and show all work.

- **While Answering Multiple-Choice (EBSR) questions.**

  - Select the bubble corresponding to your answer choice.
  - Read all of the answer choices, even if think you have found the correct answer.

- **While Answering TECR questions.**

  - Read the directions of each question. Some might ask you to drag something, others to select, and still others to highlight. Follow all instructions of the question (or questions if it is in multiple parts)

Frequently Asked Questions(FAQs)

For more information on the assessment, visit
www.lumoslearning.com/a/ilearn-faqs
OR Scan the **QR Code**

LumosLearning.com

What if I buy more than one Lumos tedBook?

Step 1 → **Visit the link given below and login to your parent/teacher account**
www.lumoslearning.com

Step 2 → <u>For Parent</u>
Click on the horizontal lines (≡) in the top right-hand corner and select **"My tedBooks"**. Place the Book Access Code and submit.

<u>For Teacher</u>
Click on "My Subscription" under the "My Account" menu in the left-hand side and select **"My tedBooks"**. Place the Book Access Code and submit.

Note: See the first page for access code.

Step 3 → **Add the new book**
To add the new book for a registered student, choose the '**Student**' button and click on submit.

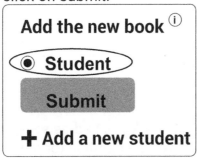

To add the new book for a new student, choose the '**Add New Student**' button and complete the student registration.

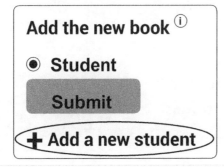

Progress Chart

Standard	Lesson	Score	Date of Completion
IAS			
3.NS.1	Read and write numbers to 1000 using base-ten numerals		
3.NS.2	Fractions of a Whole		
3.NS.3	Fractions on the Number Line		
3.NS.4 & 5	Comparing Fractions		
3.NS.6	Rounding Numbers		
3.CA.1	Addition & Subtraction		
3.CA.2	Two-Step Problems		
3.CA.3	Understanding Multiplication		
3.CA.4	Understanding Division		
3.CA.5	Multiplication & Division Facts		
3.CA.6	Multiplication & Division Properties		
3.CA.7	Applying Multiplication & Division		
3.CA.8	Number Patterns		
3.G.1	2-Dimensional Shapes		
3.G.3	Shape Partitions		
3.M.1	Liquid Volume & Mass		
3.M.2	Measuring Length		
3.M.3	Telling Time		
3.M.3	Elapsed Time		
3.M.4	Solve word problems involving money		
3.M.5	Area		
3.M.5	Relating Area to Addition & Multiplication		
3.M.6	Perimeter		
3.DA.1	2-Dimensional Shapes		

ILEARN Test Prep: Grade 3 English Language Arts Literacy (ELA) Practice Workbook and Full-length Online Assessments: Indiana Learning Evaluation Assessment Readiness Network Study Guide

Grade 3

Lumos Learning
Step Up Your Skills

INDIANA ENGLISH LANGUAGE ARTS LITERACY ILEARN Practice

Revised Edition

tedBook
ONLINE

2 Practice Tests

Personalized Study Plan

ELA Domains | Literature • Informational Text • Language

Available

- At Leading book stores
- Online www.LumosLearning.com

Made in United States
North Haven, CT
14 February 2025